Artificial Intelligence

A Non-Technical Introduction

Tad Gonsalves

Sophia University Press
上智大学出版

Artificial Intelligence

A Non-Technical Introduction

Sophia University Press

One of the fundamental ideals of Sophia University is "to embody the university's special characteristics by offering opportunities to study Christianity and Christian culture. At the same time, recognizing the diversity of thought, the university encourages academic research on a wide variety of world views."

The Sophia University Press was established to provide an independent base for the publication of scholarly research. The publications of our press are a guide to the level of research at Sophia, and one of the factors in the public evaluation of our activities.

Sophia University Press publishes books that (1) meet high academic standards; (2) are related to our university's founding spirit of Christian humanism; (3) are on important issues of interest to a broad general public; and (4) textbooks and introductions to the various academic disciplines. We publish works by individual scholars as well as the results of collaborative research projects that contribute to general cultural development and the advancement of the university.

Artificial Intelligence: A Non-Technical Introduction
© Tad Gonsalves, 2017
Published by Sophia University Press

Printed and distributed by GYOSEI Corporation, Tokyo
ISBN 978-4-324-10260-2 C3053
Inquiries: https://gyosei.jp

Dedicated to the fond memory of
my science gurus

Prof. Dr. Ludwig Boesten, S.J.
Prof. Dr. Frank Scott Howell, S.J.

who literally forced me to take up
Artificial Intelligence

and

my parents
Paulo & Isabel

who cultivated my
Emotional Intelligence

CONTENTS

CHAPTER THREE: EXPERT SYSTEMS   41

CHAPTER FOUR: FUZZY LOGIC AND FUZZY SYSTEMS 61

CHAPTER FIVE: WEB MINING AND MACHINE LEARNING 81

CHAPTER SIX: EVOLUTIONARY ALGORITHMS 101

PREFACE

Artificial Intelligence (AI) is an umbrella term spanning a history of more than half a century. Several AI disciplines have experienced a sudden surge towards the end of the last century after a prolonged period of inactivity. There is a plethora of *Artificial Intelligence* (AI) books available to the AI enthusiasts and students. What is new in this book? First of all, this book describes the efforts of the founding fathers in developing computational systems and models that put AI on a firm footing. Secondly, it covers the mainline topics in AI including a description of its latest successes at the time of writing.

The book does not try to cover all the topics in AI. Rather, it serves as a gentle introduction to the subject in non-technical terms. It introduces the reader to research topics in *Good Old-Fashioned AI* (GOFAI) as well as the latest trends in *Computational Intelligence* (CI). GOFAI aimed at the complete modelling and simulation of the human brain, while CI works with limited, but flexible systems such as Fuzzy Systems, Neural Networks and Evolutionary Algorithms that have produced promising results.

This book is primarily intended as a textbook for university undergraduate science and engineering students, with or without a computer science background. The author has been teaching AI courses for several years in the Faculty of Science & Technology, Sophia University. The contents of the book are based on the lecture notes revised and updated every year, taking into consideration the feedback provided by the students. It avoids the technical details and AI jargon, in particular. It also includes ample exercises at the end of each chapter.

This textbook can be used by universities running either a semester or a quarter curriculum. There are ten chapters in all. Since each chapter is an independent and standalone unit, readers and teachers may pick and choose chapters which suit their interests.

It is rightly said that AI is used to test the theories of the mind. The complex phenomenon, known as the *mind* cannot be sufficiently explored and researched using the tools and models developed in one particular discipline. Interdisciplinary AI borrows heavily from models and paradigms developed, deepened and used in other disciplines. Practitioners and researchers in disciplines like psychology, philosophy, and linguistics will find some valuable insights in the book.

The book offers a collection of JavaScript programs for AI enthusiasts. Most of the programs come with graphic interfaces and animations making interacting with them easier and a lot more fun. The programs are available for free download from the author's web page (http://pweb. cc.sophia.ac.jp/intellisystems/).

Finally, I wish to extend a word of gratitude to my friends and well-wishers who couldn't wait to see the birth of my first AI book, and especially to Profs. Bob Deiters and Mike Milward, for painstakingly checking and re-checking the manuscript. Thanks to one and all!

Tad Gonsalves, Ph.D.

CHAPTER ONE
SILICON INTELLIGENCE

1.1 Introduction

The English mathematician Charles Babbage, who designed the Analytical Engine in 1837, is considered to be the inventor of modern computers. However, the first link between machines and intelligence was proposed by Alan Turing in 1950 in his paper entitled, "Computing Machinery and Intelligence." According to the *Turing Test* proposed in this paper, a computer can be judged to be intelligent if the responses given by it to our queries are indistinguishable from those given by a human person.

The term *Artificial Intelligence* (AI) was coined by John McCarthy in a conference held in Dartmouth, New Hampshire in 1956. He defined AI as "the science and engineering of making intelligent machines, especially intelligent computer programs." Being an eminent computer sceintist, he considered intelligence as the *computational part* of the ability to achieve goals in the world.

In the opening chapter of this book, we shall attempt to clarify the concepts that lead to a sound definition of Artificial Intelligence. We shall also acquaint ourselves with the short history of AI, current research trends, and its future promises.

1.2 Defining Artificial Intelligence

The definition of Artificial Intelligence rests on two important concepts: *Artificial* and *Intelligence*. Therefore, we may reframe our question in two separate questions: (1) What is the meaning of *Artificial*? (2) What is the meaning of *Intelligence*? If we can clearly define each of these terms independently, our task of defining AI will be successful.

1.2.1 What is being artificial?

Our first impression of something artificial is that it looks like the real or natural thing. But there is another more important aspect to it; namely, it performs a *function* similar to that of the natural thing that it resembles. Take, for example, artificial legs. They resemble natural legs in form and function; they look like natural legs and they help a person to walk. Some artificial objects might just have a resemblance without the function, like artificial eyes, for example (Fig. 1.1); they look like eyes, but they cannot see (although *Bionics* may change this in the future).

We can stretch our imagination a bit further and think of a *function* that is artificial, while the appearance may be totally different from the natural thing (Table 1.1). Thus, we may say that a ship does *artificial swimming* or an airplane does *artificial flying*.

Ships and airplanes are, respectively, far-fetched imitations of fish and birds, but their functions closely resemble those of fish and birds. However, even these functions, though similar, are far-fetched. A ship is *swimming* and an airplane is *flying*; but we do not see the ship maneuvering through the water with some fin-like apparatus nor do we expect the airplane to fly by fluttering its wings.

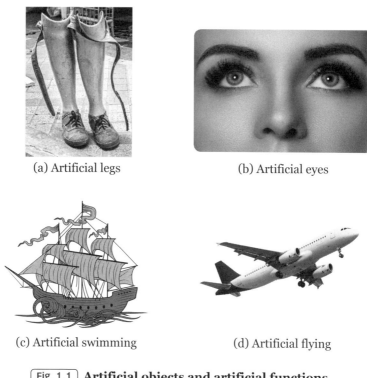

(a) Artificial legs   (b) Artificial eyes

(c) Artificial swimming   (d) Artificial flying

Fig. 1.1 **Artificial objects and artificial functions**

Artificial Intelligence is somewhat similar to ships swimming and airplanes flying. To complicate matters, intelligence is not an object like eyes, legs, ships, or airplanes. Artificial Intelligence neither resembles the form nor the function of natural intelligence. When we say a ship is sailing, we do not imagine it to be moving in the ocean like a dolphin or a whale. When we say that an airplane is flying, we do not imagine it flapping its wings. When we say that a person with artificial legs is walking, we do not imagine blood circulating and nerve messages passing through the legs. But we look at the outcome that is very similar to walking. Similarly, when we say a machine is *thinking* we may not imagine that there are thoughts running inside the core of the machine or that the machine is *conscious* of its own computations. Computers, or more precisely AI programs, do not think while scratching their heads as do human beings. It is difficult to imagine AI because it resembles natural intelligence neither in form nor in function.

Table 1.1 **Artificial objects: their appearance and function**

Artificial objects	Appearance	Function
legs	✓	✓
eyes	✓	×
ships	×	✓
airplanes	×	✓

By extending our metaphor, we may think of humanoid robots as *artificial* human beings. We cannot and do not refer to their internal mental states. Rather, by looking at their behavior and actions, we infer that they somehow manipulate knowledge in an intelligent way to solve a given problem.

1.2.2 What is intelligence?

Intelligence is a highly complex phenomenon that cannot be easily defined. For the sake of simplicity and clarity, we shall adopt the definition of intelligence found in the Webster's dictionary: "*Intelligence is the ability to learn and solve problems.*" Our aim is not to use this definition as a criterion to classify living creatures into intelligent v/s non-intelligent categories, but as a working definition to explain and explore the various AI models covered in this book.

When hungry, almost all animals "solve" the problem of hunger by looking for food and eating it when found. Up to this point, they rely on their survival instincts (their genetic hard-wired program). Nevertheless, quite a few species of animals will give up the search if the food is not within reach and go looking elsewhere. This implies they cannot do more than what is genetically programmed into them. They cannot modify the hard-wired program in their brain; in other words, they cannot *learn*. According to the above definition, subjects not capable of learning to adjust to unexpected situations fail to make it to the intelligent class.

Fig. 1.2 **Monkey displaying self-learnt skills**

A monkey, on the other hand, demonstrates the ability to adapt its behavior to changing conditions. It will start looking for food when hungry, acting on its instincts. However, when the food is unreachable, it will devise means to get the food within reach. For instance, it will get hold of a stick to pull fruits or nuts closer. It will also peel the skin off the fruits or crack the nuts before eating (Fig. 1.2). Moreover, with every trial, it will improve its skills. The monkey's behavior of adapting to changing situations and learning to solve problems is a sign of intelligence.

Fig. 1.3 **Dolphins displaying human-taught skills**

Dolphins have natural hunting and social skills. In addition, they can be trained to acquire magificent entertaining skills (Fig. 1.3), which further demonstrates their innate capacity for learning. No wonder, their brain-to-body mass ratio (or encephalization quotient), which is a rough measure of the intellgience in animals, is second only to that of humans.

Living beings do not qualify to be called intelligent if they do no more than execute the hard-wired program in their brains or DNA. Similarly, no matter how powerful and speedy a computer may be, if it can do no more than what is dictated by the programed instructions inside its memory, it hardly qualifies to be called intelligent. Human beings have the highest kind of intelligence which enables them to learn and solve problems. The aim of AI is to imitate this intelligence in machines. Therefore, AI may be defined as:

- The science of making machines that think and behave like human beings.
- The science of making machines do the things which, at least at present, can be done only by human beings.

We shall adopt this definition of Artificial Intelligence as we go through the various chapters of this book. It will serve as a guide for an in-depth study of the contents of this book.

1.3 Kinds of AI

In this section, we shall discuss the different kinds of AI, like weak AI, strong AI, and extended AI.

1.3.1 Weak AI

An Expert System (discussed in chapter 3), is a good example of a successful weak AI. It mimics the knowledge and expertise of experts in a given field (domain) and gives advice to the end users. On an average, the performance of Expert Systems is found to be on a par with that of experts, at times even excelling them. Chess playing super-computers are also examples of weak AI. They perform at the grandmaster level and often beat human grandmasters. Weak AI excels at the task for which it is designed. However, its knowledge and expertise usually cannot be extrapolated to solve problems in other domains. Needless to say, it does not have the higher human faculties such as thinking, feeling, imagining, etc. It does not have any common sense.

1.3.2 Strong AI

The goal of strong AI is to create machines with intellectual and "spiritual" capabilities no less than those of *Homo Sapiens Sapiens*! Strong AI machines will have the general knowledge and ability to ponder and think over a problem, and acquire sufficient skills to solve the problem. Possessing emotions and will, they will also interact with human beings. They will also possess that salient ability which has been unique only to human beings: They *will know that they know*. Yes, they will have consciousness just like us.

1.3.3 Extended AI

The original aim of the AI founding fathers was to create armless robots (and we hope "harmless," too!); that is, computer programs that imitate the knowledge and expertise of human beings in solving theoretical and practical problems. They were not concerned with the design of dexterous limbs and other bodily functions of human beings. Therefore, the Turing Test (explained in detail in chapter 10) is offered in an environment without sophisticated input-output devices. Traditional input-output devices like a mouse, keyboard, and screen are sufficient to demonstrate the intelligence of a program residing inside the memory of a computer. What about *Robot-*

ics, then? The science and engineering devoted to the design of robots, especially humanoid robots that *look like and behave like human beings*, may be called extended AI.

1.4 Brief history of AI

The early programs for proving theorems and playing checkers astonished the public with their seemingly intelligent performance. Expert Systems, too, drew great admiration since their performance in consultation was on a par with human experts. However, the expectations raised by these early systems led to the disappointment of AI in the ensuing years. It is only at the turn of the century that AI began showing signs of coming back to life and progressing towards a promising future. This section briefly describes the tumultuous history of AI.

1.4.1 The hype cycle of AI

The hype cycle[1] is a graphical representation of the stages a technology goes through from its conception to maturity. As happens to any new technology, the history of AI, too, ran through a hype curve. The early AI programs like the ones elegantly proving theorems and skillfully playing board games aroused great interest and expectations. This was followed by the successful application of Expert Systems in business and academia. This early period in the development of AI is referred to as the "peak of inflated expectations"

---

1   http://www.gartner.com/technology/home.jsp (accessed in March, 2017).

shown in the AI hype curve (Fig. 1.4).

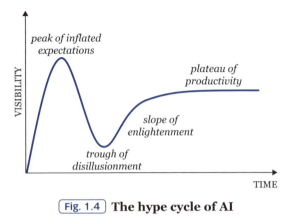

Fig. 1.4 The hype cycle of AI

However, for the next ten years AI did not live up to its expectations. Not getting substantial returns, the AI stakeholders appointed
special committees to examine the progress of AI. This period,
called the "trough of disillusionment," proved harmful to the development of AI. Funds were heavily cut and AI research almost came
to a grinding halt. Despite these setbacks, some AI researchers per-
sisted with their research in a few selected areas, refraining from
calling their work "AI." Their approach finally led to the "slope of
enlightenment," bringing AI out of the harsh winters. From the
1990s, AI research began to flower again making breakthroughs in
Machine Learning and Game Playing. By the end of the last century,
AI had reached the "plateau of productivity."

1.4.2 The heydays of Expert Systems

An Expert System is a computer program created to solve problems at a level comparable to that of a human expert in a given domain. Also known as Knowledge-based Systems, Expert Systems rely on expert human knowledge built into a computer program to solve problems. The system has a vast store of knowledge in the knowledge-base and uses some kind of reasoning process to infer solutions to a given problem. The problems that can be solved by an Expert System are restricted to its particular domain. Accordingly, Expert Systems do not display general intelligence like human beings, but can be extraordinarily skillful in analyzing and solving problems in specialized domains. The following are two well-known examples of successful Expert Systems:

DENDRAL

DENDRAL (acronym coined from DENDRitic ALgorithm) was the earliest Expert System developed in 1965 at Stanford University by a team of computer scientists, organic chemists and geneticists. It was a chemical-analysis Expert System programmed in LISP, the most prominent AI language of that period. The substance to be analyzed had its spectrographic data fed into the system, and the system would hypothesize the substance's molecular structure. For example, the mass spectrum of water molecule shows a peak at 18 units. This is because the water molecule (H_2O) consists of two molecules of hydrogen (H) and one molecule of oxygen (O) and their molecular weights are 1 and 16, respectively. DENDRAL would refer to the atomic mass numbers and valence of elements to determine the constituents whose mass would add up to 18.

DENDRAL's performance rivaled that of an expert chemist. However, this was not because the program knew more than an expert, but because of its systematic search through the space of possibilities and its systematic use of what it knew. It was a convergence of three different technologies: Knowledge Engineering, Machine Learning and Heuristics Programming.

MYCIN

MYCIN was a medical Expert System developed at Stanford University in the early 1970s. Its task was to diagnose infectious blood diseases and recommend antibiotics as a doctor in real-life would do. Although the inferences were based on a knowledge-base containing a modest 500 rules, MYCIN's success was impressive. Besides diagnosing bacterial infections and prescribing medication, it would also give an explanation at the end of its reasoning procedure to the patient, thereby increasing its credibility.

A typical rule in MYCIN has an antecedent and a consequent. We refer to the antecedent of a rule as the premise and to the consequent as the action. The inference can either take a forward chaining (data-directed) or a backward chaining (goal-directed) mechanism (chapter 3). MYCIN primarily used a backward chaining control strategy. MYCIN was found to perform better than the average physician in diagnosis and prescription. However, due to legal and ethical issues, it was never used commercially.

1.4.3 AI misfortunes

AI, in its long history has had many setbacks. Some of them are briefly discussed in the following sub-sections.

Connectionism on the decline

The field of Artificial Neural Networks (ANN), also known as Connectionism, is currently a budding field of AI. However, the technology was nipped in the bud in the 1960s. The basic neuron, called a perceptron, had weights assigned to the inputs and learning consisted of adjusting the connection weights to produce appropriate outputs.

Perceptrons, in general, serve as linear threshold functions. It was pointed out that the perceptron could only solve linearly separable functions, but could not classify more complex data. Criticism by rival researchers led to the decline in perceptron research until about the 1980s. The limitations of the perceptron model were overcome by creating neural networks with hidden layers. These networks connect the neurons of the input layer to those in the output layer via the neurons in a hidden layer and correctly classify the input data.

Two fatal reports

In the 1960s and 70s, referred to as the "winter of AI," research suffered drastic funding reduction that led to its downfall. Two major reports were responsible for this downfall. The first was the ALPAC

Report published in the US in 1960; the second, the Lighthill Report published in the UK in 1973.

The ALPAC Report

In the 1950s, many universities in the US, UK and Russia started major machine translation projects. At Georgetown University, in particular, researchers began building a pilot system to convince potential funding agencies of the feasibility and utility of machine translation. In 1954, this led to the famous Georgetown experiment, a pilot system translating from Russian to English, which was a grand success. During the next ten years, various US agencies invested millions of dollars in machine translation.

However, soon the AI researchers faced grand challenges in natural language processing and in machine translation, in particular. In 1964, owing to the lack of appreciable results, the US government appointed a committee to evaluate the machine translation research. This "Automatic Language Processing Advisory Committee" (ALPAC), compiled a report in 1966 which proved fatal to machine translation research. The report stated that machine translation besides being slow and inaccurate was more expensive than human translation. Machine translation research funds were soon suspended and for years after that it was portrayed in bad light in public media.

The Lighthill Report

The Lighthill Report compiled in the UK in 1973, provoked a great loss of confidence in AI which the academics and funding bodies used to have. It pointed out three main drawbacks of AI, calling all AI achievements "past disappointments."

1. Automatic aircraft landing system

   According to the Lighthill Report, the conventional technology using radio waves for aircraft landing was more successful than the costly AI automatic system which it claimed was not practical.

2. Weakness of AI chess programs

   According to the Lighthill Report, the sophistication of the AI chess programs only reached that of an "experienced amateur" at that time. They were far from beating a grandmaster.

3. Combinatorial explosion problem

   The Lighthill Report pointed out that all AI methods required substantial knowledge of the subject matter in order to be useful. Since the AI knowledge acquisition was not automatic, the report maintained that AI would not be able to keep up with the combinatorial explosion problem.

1.4.4 The triumphs of AI

AI had three major successes, beginning in the last decade of the 20th century. These three major successes, chiefly in game-playing, are discussed at length in chapter 8. They are briefly mentioned here.

AI wins world-class chess

In 1997, the IBM supercomputer *Deep Blue* defeated the reigning chess world champion, Gary Kasparov. The supercomputer was especially designed to play chess. It could examine about 200 million moves per second. It also had a large database containing the games played by chess grandmasters. *Deep Blue*'s success was due to the advanced game playing AI algorithms.

AI wins world-class Jeopardy game

In 2011, IBM's supercomputer *Watson* defeated the reigning Jeopardy champions, Brad Rutter and Ken Jennings. Jeopardy is a well-known TV quiz in the US. It consists not in answering questions, but rather framing questions when given a wide body of related hints. *Watson* was designed to take the hints from the Quiz, do natural language processing and frame the Jeopardy questions. A Jeopardy question can be complex because the contestant has to put a large number of facts together in a very short time.

AI gives it a *Go*

In March 2016, DeepMind's AI *Go*-playing supercomputer *AlphaGo* defeated Korean grandmaster Lee Sedol, winning 4 out of 5 games in series. *AlphaGo* trained by Machine Learning algorithms mastered the 2,500-year-old *Go* game that is exceedingly more complex than chess and requires a great deal of intuition and creativity. Many experts remarked that in the widely telecast championship match, *AlphaGo* displayed an unorthodox style of playing.

1.4.5 Current trends in AI

Where does AI stand today after going through periods of bleak winters and dry summers? Has it seen any spring? Yes. Scratch the surface of any computer or smartphone and you will see dozens of AI programs working together to give you sleek applications. Our daily Web search and browsing, social networks and the host of apps we could not do without are all driven by AI programs.

It is said that the computer entertainment industry generates more income than Hollywood. Video games rely on computer vision and AI planning. Healthcare to elderly people is another beneficial application of AI in the form of robotics. In education, personalized AI tutors coach students in math, science, programming and learning natural languages. AI security systems are always vigilant and are constantly battling against the cybercrimes perpetrated by humans.

The greatest success that AI has shown in recent years is in the area of deep learning (more about this in chapter 5). Applications in image and face recognition, speech recognition and synthesis, natural language comprehension and translation, self-driving cars and a host of other unforeseen apps are on the horizon.

It is true that mass media shape most people's imagination about the frightening aspects of AI. Needless to say, AI in the movies is not real AI. It is science fiction AI. Many experts still argue that building computers smarter than humans is an unnecessary and futile effort. However, the aim of the science of AI is not to make computers smarter than humans, but to make them smart enough to help us make our lives more comfortable, safer, healthier, and longer.

We are constantly pushing the limits of technology to build machines to enhance our physical and mental faculties. And we are very proud of our achievements. It is a strange fact that no world-famous sprinter ever feels inferior to a *man-made* Ferrari covering 100 meters in less than 6 seconds, no Olympic weightlifting champion to a *man-made* bulldozer lifting mega-tons of weights, and no accountant to a *man-made* pocket calculator performing 10-digit calculations in the fraction of a second. Yet, almost all of us would feel inferior to a successful *man-made* silicon intelligence!

1.5 Exercises

1. *Machine Intelligence*, the term suggested by Alan Turing is more appropriate than *Artificial Intelligence*, since it does not connote any comparison to human natural intelligence. What are some of the more accurate terms for *Artificial Intelligence*?

2. Your spam filter regularly checks for unwanted mails and puts them away. Whenever you miss-spell a word in Google search, the search engine corrects it instantaneously for you. There are countless game playing programs, playing at different levels of difficulty, which you can play against free of charge. Are these programs *intelligent*?

3. Humans are known for their creativity, which is expressed in the form of poetry, literature, art, music, etc. Do you think future AI computers will be *creative*? To what degree?

4. Design an experiment or a method to test the intelligence of computers. Is the IQ test reliable?

5. What are some of the pertinent *moral* issues computer scientists and engineers face when developing AI?

6. Will the future AI be for or against humanity?

7. The strong AI protagonists are convinced we need only further advances in technology to make strong AI possible. The antagonists, on the other hand, maintain that no amount of development in technology can ever lead to the development of strong AI, because humans contain a non-material faculty called the soul. Which philosophical position will you hold? State your reasons.

8. Granted that human intelligence cannot, in principle, be imitated in machines due to profound philosophical reasons. Is artificial *animal* intelligence possible? In other words, can we create intelligent cats, dogs, or dolphins that act and behave like their natural counterparts?

9. Thought Experiment 1: Imagine all your brain functions like thinking, reasoning, remembering and feeling are downloaded to a silicon brain. Will that brain re-produce your identity?

10. Thought Experiment 2: Imagine a scenario in the year 2050. The person sitting beside you in the train suddenly confesses that although she looks exactly like a human being, she is in

fact a machine fully conscious of herself. How will you refute her claims?

References

Buchanan, B. G., & Shortliffe, E. H. (Eds.). (1984). *Rule-Based Expert Systems: The MYCIN Experiments of the Stanford Heuristic Programming Project*. MA: Addison-Wesley.

Crevier, D. (1993). *AI: The Tumultuous history of the search for artificial intelligence*. New York: Basic Books.

Lindsay, R. K., Buchanan, B. G., Feigenbaum, E. A., & Lederberg, J. (1980). *Applications of Artificial Intelligence for Organic Chemistry*: The Dendral Project. McGraw-Hill Book Company.

Luger, G. F. (2008). *Artificial Intelligence: Structures and Strategies for Complex Problem Solving*. (6th ed.). Pearson.

Lungarella, M. (2007). *50 years of artificial intelligence essays dedicated to the 50th anniversary of artificial intelligence*. Berlin: Springer.

McCarthy, J. (2001). *What is Artificial Intelligence?* Stanford University.

Nilsson, N. J. (2009). *Artificial Intelligence: A New Synthesis* (1st ed.). Elsevier.

Russell, S., & Norvig, P. (2016). *Artificial Intelligence: A Modern Approach*. (3rd ed.). Pearson.

Schalkoff, R. (2011). *Intelligent Systems: Principles, paradigms, and pragmatics*. Sudbury, MA: Jones and Bartlett.

Tim Jones, M. (2009). *Artificial Intelligence: A Systems Approach*. Sudbury, MA: Jones and Bartlett.

Warwick, K. (2013). *Artificial Intelligence: The Basics*. Routledge.

CHAPTER TWO
SEARCHING THROUGH A HAYSTACK

2.1 Introduction

One of the easiest methods of solving a scientific problem, we are taught in school, is by plugging numerical values in the proven formula. Provided we do not make a mistake in the hand calculations, we are sure to come up with the correct solution in a couple of minutes. For more complex problems, we transpose the same problem-solving methodology to our electronic calculators or computers.

There is another simple problem solving method known as *search*. As the name implies, you are given a *search space* to conduct your search. The solution already exists in the search space. All you have to do is find an intelligent search strategy. We are accustomed to think that computers can search through a haystack of information in the fraction of a second. However, given a simple problem with sufficiently large number of search points, the time needed even for a supercomputer to search through all the possible combinations of the search points would be more than the age of the universe!

In this chapter, we shall first discuss about general algorithms to solve problems, pseudocode, and code structures and then look at some of the well-known search algorithms like the Best First Search and the A* (A star) algorithms.

2.2 Algorithms

In order to solve a problem, the computer needs a clear and detailed set of steps. This clear, unambiguous, and detailed set of steps to solve a problem is called an *algorithm*. A computer program is a detailed set of instructions coded in a particular computer language, that follows the algorithm. Computer programming languages, as explained in this section, have well-defined structures to code the algorithms.

2.2.1 Pseudocode and program

The detailed steps to solve a problem is provided by the algorithm. The algorithm is then converted into a *pseudocode*, which is a mixture of natural language and computer language. The pseudocode is then converted into a *code*, which is pure computer language. For simple problems, one writes the program code directly without going through the intermediate steps of explicitly writing down the algorithm in the form of a pseudocode. However, for complex problems, the pseudocode-to-code procedure must be followed to avoid mistakes. Fig. 2.1 describes the algorithm to determine if a natural number n is even or odd; Fig. 2.2 gives the pseudocode, and Fig. 2.3 the actual code in the Java programming language.

Divide n by 2

If the remainder is 0, n is *even*

If the remainder is 1, n is *odd*

Fig. 2.1 **Algorithm to determine if n is even or odd**

Start

Input number n

If (n MOD 2 = 0) Then

      display "Number is even"

   Else

      display "Number is odd"

End If

Stop

Fig. 2.2 **Pseudocode to determine if n is even or odd**

```java
import java.io.IOException;
import java.util.Scanner;
public class EvenOddNumber {
  public static void main(String[] args) throws IOException {
    Scanner s = new Scanner(System.in);
    int n = s.nextInt();
      if (n % 2 == 1)
        System.out.println("n is odd");
      else
        System.out.println("n is even");
} }
```

Fig. 2.3 **Code to determine if n is even or odd**

2.2.2 Basic program control structures

Normally, a program executes steps sequentially from top to bottom. The program is stored in the main memory of the computer. The Central Processing Unit (CPU), the brain of the computer, fetches one instruction at a time from the main memory, decodes and executes it. Program execution by the CPU is a simple repetitive *fetch-decode-execute* cycle. What makes this impressive is the extraordinary speed of execution (millions of cycles per second!).

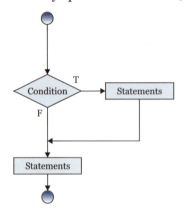

Sometimes the top-bottom sequential flow is interrupted to test a condition. If the test result is true, the program follows another sequence of commands. If not, it continues on the original sequential path (Fig. 2.4).

Fig. 2.4 **Program control structure:** *if*

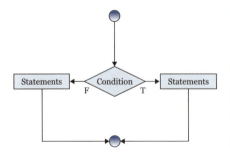

In the *if-else* control structure, the condition is first evaluated. If the condition is true, the program follows the statements in the T direction. If the condition is false, it follows the statements in the F direction (Fig. 2.5).

Fig. 2.5 **Program control structure:** *if-else*

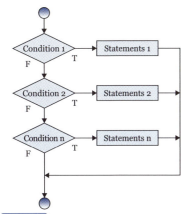

Sometimes, the program categorizes the values of a variable, for instance, in grading examination results. This is efficiently done by the multiple *if else-if else* structure (Fig. 2.6).

Fig. 2.6 **Program control structure: *if else-if else***

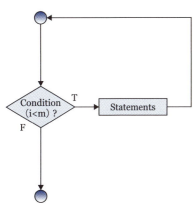

Very often, a number of computations have to be repeated, as when grading the results of an exam taken by a hundred students. In this case, the *for* loop is repeated a hundred times (Fig. 2.7).

Fig. 2.7 **Program control structure: *for loop***

2.3 Solving search problems

In this section, we shall first define a search problem in very simple non-mathematical terms, and then describe some classic search algorithms.

2.3.1 Search problem

A search problem contains a set of states or nodes {a, b, c, d, e, ...}.

An action or a move induces a change in state. Given a start node and a goal node, the search problem is to find a sequence of nodes that leads from the start node to the goal node. For example, on your first visit to a city, you may wish to walk from the station (starting node) to the market (goal node). With a tourist map in hand, you may walk from the station to the obelisk, then to the theater, park, coffee shop, book store, and finally to the market, assuming that to be the shortest route. The solution to this search problem is: station, obelisk, theater, park, coffee shop, book store, market. A similar strategy is followed by AI game playing algorithms as we shall see later in this chapter.

2.3.2 Search space

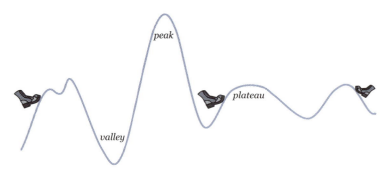

Fig. 2.8 **Search landscape**

The search space or the solution space is a set containing all the solutions to the search problem. The problem of finding a solution in the search space is often compared to hill climbing in a given terrain or landscape. The hill-climbing metaphor covers terms such as terrain, plateaus, peaks, valleys, etc. (Fig. 2.8). Further, we make use of two important mathematical concepts called *graph* and *tree* to construct a search algorithm. They are explained below.

Graph

A graph, in mathematics, is a set of nodes and arcs. Arcs or edges connect the nodes. At times, arcs have directionality indicated by arrows. In Fig. 2.9, the nodes are : {p, q, r, s, t, u}; the arcs are: {(p,q), (p,u), (q,r), (u,t), (r,s), ...}.

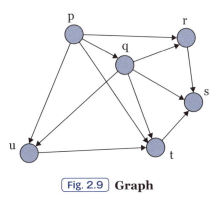

Fig. 2.9 **Graph**

The nodes represent a set of states in a problem, while the arcs represent actions or moves that change the state from one node to another. Arcs can be directed (indicated by arrows) or non-directed. A sequence of nodes through successive arcs constitutes a path. The solution to a search problem is a path. (Recall that the solution to the problem of going from the station to the market with a tourist map in hand is essentially finding a *path*).

Tree

A tree is a special case of a graph in which two nodes have no more than one path between them. The starting node is called the root and the end nodes are called leaves. A node is called parent node if it has children nodes connected to it. Multiple nodes proceeding from the same parent are called siblings. The arc or edge between two nodes is called a branch. The tree structure resembles a natural tree turned upside down (Fig. 2.10).

Armed with a graph and/or tree, one can find the solution using any of the search strategies described below.

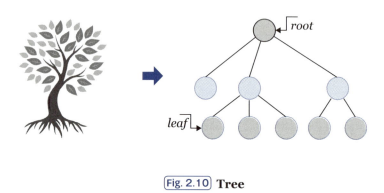

Fig. 2.10 **Tree**

2.3.3 Uninformed search

When no hint or clue about the reachable path is available, the obvious thing to do is to search through all possible paths. This strategy is called by various names: uninformed search, blind search, exhaustive search or brute force. First, the search tree is constructed by referring to the actual nodes and the connecting arcs. The Breadth First Search algorithm traverses the search tree along its breadth (Fig. 2.11), while the Depth First Search algorithm traverses the search tree along its depth (Fig. 2.12).

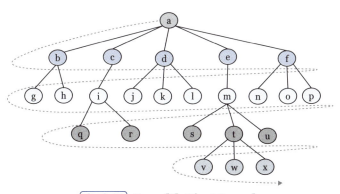

(Fig. 2.11) **Breadth First Search**

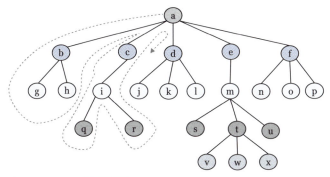

(Fig. 2.12) **Depth First Search**

An uninformed search has several limitations. It can easily get trapped on plateaus and local optima. The greatest weakness of an uninformed search, however, is its brute-force nature of trying to explore all possible paths. The brute-force method hardly deserves to be called intelligent.

2.3.4 Informed search

Informed search is also called heuristic search: a *heuristic* refers to a rule of thumb to solve a problem in a given domain. It relies on partial knowledge of the structure or the features of the search space to conduct the search. In search problems, a heuristic is an evaluation function whose value serves as a rough estimate of the cost of reaching from a given node to the goal node. In computer science, *cost* can mean the actual monetary cost, time, distance, effort or any other suitable metric.

Best First Search (BFS) and A* (pronounced, "A star") are two well-known heuristic search algorithms, each of which is described below.

Best First Search

The evaluation function or the heuristic h(n) is the estimate of cost from n to the closest goal. In a road navigation problem, h(n) is the straight-line distance (SLD) from the current location n to the goal location. The BFS algorithm is given in Fig. 2.13.

Make two lists: open list (O list) and closed list (C list)

Place the start node in the O list

The C list is initially empty

while (O list is not empty) {

    S is the first node in the O list

    place node S in the C list

      if (S is the goal node) {

          return the search path

    }

    else {

        place the not-yet-searched children of S in the O list

        arrange the O list in ascending order of h values

    }

}

Fig. 2.13 **Best First Search pseudocode**

Finding the shortest route using BFS

To find the shortest path using BFS from node A to node E in Fig. 2.14, we refer to the SLD heuristic values from node n to the final node E tabulated in Table 2.1. (Note that infinity signifies dead ends.) We follow the BFS algorithm given in Fig. 2.13 and expand the search tree shown in Fig. 2.15.

Table 2.1 Straight line distance (SLD) from node n to E

Node	SLD to E	Node	SLD to E
A	159	M	117
B	43	N	108
C	66	O	170
D	105	P	69
E	0	Q	∞
F	96	R	110
G	∞	S	136
H	21	T	149
I	85	U	43
J	∞	V	59
K	134	W	∞
L	134	Z	168

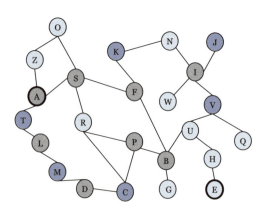

Fig. 2.14 Finding the shortest route from node A to E

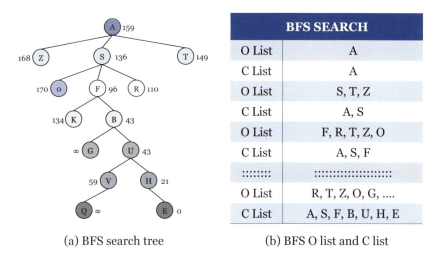

	BFS SEARCH
O List	A
C List	A
O List	S, T, Z
C List	A, S
O List	F, R, T, Z, O
C List	A, S, F
::::::::	:::::::::::::::::::::::
O List	R, T, Z, O, G, ...
C List	A, S, F, B, U, H, E

(a) BFS search tree   (b) BFS O list and C list

Fig. 2.15 **Finding the shortest route using BFS**

Solving the 8-puzzle using BFS

The 8-puzzle consists of 8 tiles numbered from 1 to 8, placed in a 3 x 3 slot frame. Since there are only 8 tiles, one of the slots is left blank. The tiles in the vicinity of the empty slot can slide into the empty slot. The sliding action of a neighboring tile into the empty slot is called a *move*.

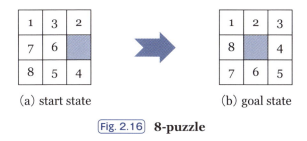

(a) start state   (b) goal state

Fig. 2.16 **8-puzzle**

The ordering of the tiles clockwise from 1 to 8 along the edge of the puzzle board with the blank slot in the middle represents the goal state. The start state of the puzzle is a jumbled state in which at least one of the tiles is out of place compared to the goal state (Fig. 2.16).

The constraints of the game may be framed as follows: the tiles lying on the edge of the game board may not be moved beyond the edge. For example, 1, 3, 2 may not be moved up; 1, 7, 8 may not be moved left; 8, 5, 4 may not be moved down; 2, 4 may not be moved right. Adjacent tiles may not be moved against each other. For example, 1 may not be moved down or right; 3 may not be moved left, right or down, etc.

Recall that solving the 8-puzzle consists in finding a sequence of moves carrying the puzzle from its initial state to the goal state. The BFS algorithm can be used to solve the 8-puzzle problem just as we have solved the path finding problem. In order to apply the BFS, let us define the necessary parameters as follows:

- Node: A configuration of the tiles on the puzzle board
- Move: Sliding a tile into an empty slot
- Heuristic function: Sum of the Manhattan distance of the tiles which are out of place

$h = 4$ $h = 0$

Fig. 2.17 **Initial and goal state of the 8-puzzle**

Manhattan distance (MD) or city block distance counts the number of blocks a tile has shifted from its goal position. The heuristic function is defined as:

$$h = \sum_{k=1}^{8} MD(pos_s^k, pos_g^k)$$ (2.1)

where, pos_s^k and pos_g^k respectively represent the start and the goal position of the kth tile in the puzzle. Table 2.2 shows the heuristic values of all the tiles in the starting state (initial configuration).

Table 2.2 **Heuristic values of the tiles (nodes)**

Heuristic	TILES								Total
	1	2	3	4	5	6	7	8	
h (start)	0	0	0	1	1	2	0	0	4
h (goal)	0	0	0	0	0	0	0	0	0

The BFS starting node is the puzzle configuration shown in Fig. 2.17. The empty space can be filled in four different ways, giving rise to four children. Their heuristic values are found to be 4, 3, 5, and 3. The node with the lowest heuristic value is chosen as the next move. When there is a tie between two or more nodes, as in this case, any one can be chosen. The BFS algorithm proceeds till the end node is reached (Fig. 2.18).

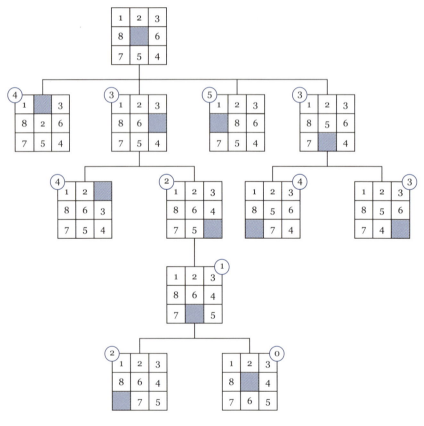

Fig. 2.18 BFS solution to the 8-puzzle

A* Algorithm

The A* (A star) algorithm is a well-known path-finding algorithm, extensively used in developing computer games. The algorithm is shown in Fig. 2.19.

1. Name the starting point S
2. Calculate the h value of S
3. Place S on the O list
   Initially, S is the only node on the O list
4. Let B = the lowest f-value node on the O list
   If B is the goal node, then end search (path found)
   If the O list is empty, then end search (path not found)
5. Let C = child node of B
   Calculate the h value of C
   Check whether C is on the O list or C list
      If so, check whether the new path has lower f-value
      If so, update the path
      Else, add C to the O list
   Repeat step 5 for all the children of B
6. Repeat from step 4

Fig. 2.19 **A* algorithm pseudocode**

What distinguishes A* from other algorithms is that it takes into account the cost of reaching the current node from the start node in the objective function. The total search cost, f(n) is given by:

$$f(n) = g(n) + h(n) \qquad (2.2)$$

where, g(n) is the actual lowest cost of the path from start node to node n and h(n) is the estimated cost from node n to the goal node.

2.3.5 Meta-heuristic search

Meta-heuristic search is the most general and most efficient way of solving a search problem, specially an NP-hard problem because it does not rely on any domain heuristics. The computational cost does not increase appreciably with the size of the problem. Most of the meta-heuristic algorithms are nature-based or bio-inspired. These algorithms are discussed at length in chapters 6 and 7. The Genetic Algorithm is a typical meta-heuristic algorithm. It starts with a random set of solutions called a population and computationally mimics the process of biological evolution to evolve them. Refer to chapter 6 for the explanation of solving difficult search problems using the Genetic Algorithm.

2.4 Exercises

1. Why is it advisable to write an algorithm as a pseudocode, rather than directly as code?

2. Write down a pseudocode for washing your clothes.

3. Think of an efficient algorithm to find the telephone number of one of your friends in the telephone directory. Write down the algorithm in words and then convert it into a pseudocode.

4. The sieve of Eratosthenes is a classic algorithm to find the prime numbers up to a given integer N. Describe the algorithm in words. Write a pseudocode from the above description.

5. Imagine you are travelling to a city for the first time. You want to visit a famous museum in the city. You go to the tourist information center and request a map on which you see a hospital, library, church, temple, garden, restaurant, monument, city office, theater and the museum. Sketch the map and estimate the straight line distances from the tourist information center (your start node) to the museum (your end node). Use the BFS algorithm to find a path to your goal.

6. Use the A* algorithm to find the optimal path in the above problem.

7. Some of the often-used distance metrics are: Euclidean (as-the-crow-flies), Manhattan (city-block), and Hamming. With suitable examples explain how to measure each of these distances.

8. Draw a maze and use the A* algorithm to get out of the maze from an arbitrary starting position.

9. With h = number of tiles out of place, solve the 8-puzzle using BFS.

10. As of now, human researchers develop algorithms and program computers to execute them. Can AI programs invent algorithms?

References

Agnarsson, G., & Greenlaw, R. (2007). *Graph theory: Modeling, applications, and algorithms.* Upper Saddle River, NJ: Pearson/ Prentice Hall.

Allman, J. (2012). *Algorithms.* Niantic, CT: Quale Press.

Christian, B., & Griffiths, T. (2016). *Algorithms to Live By: The Comput-*

er *Science of Human Decisions*. Henry Holt and Co.

Cormen, T. H., Leiserson, C. E., & Rivest, R. L. (1990). *Introduction to algorithms*. Cambridge, MA: MIT Press.

Dasgupta, S., Papadimitriou, C. H., & Vazirani, U. V. (2008). *Algorithms*. Boston: McGraw-Hill Higher Education.

Edelkamp, S., & Schroedl, S. (2011). *Heuristic Search: Theory and Applications*. Morgan Kaufmann.

Heineman, G. T., Selkow, S., & Pollice, G. (2009). *Algorithms in a nutshell*. Sebastopol, CA: O'Reilly.

Johnsonbaugh, R., & Schaefer, M. (2003). *Algorithms*. Upper Saddle River, NJ: Pearson Education.

Karumanchi, N. (2016). *Data Structures and Algorithms Made Easy: Data Structures and Algorithmic Puzzles* (5th ed.). Career-Monk Publications.

Knuth, D. E. (1997). *The art of computer programming. Sorting and Searching*. Reading, MA: Addison-Wesley.

Knuth, D. E. (2011). *The art of computer programming*. Reading, MA: Addison-Wesley.

Pearl, J. (1984). *Heuristics: Intelligent Search Strategies for Computer Problem Solving*. Reading, MA: Addison-Wesley.

Skiena, S. S. (2009). *The Algorithm Design Manual* (2nd ed.). London: Springer.

CHAPTER THREE
EXPERT SYSTEMS

3.1 Introduction

The human mind is capable of storing vast amounts of knowledge. It also has the remarkable ability to organize knowledge, and combine it with relevant pieces of data to solve problems. Building a machine with the capability and flexibility close to that of the human mind is, at least at present, next to impossible. However, we can divide knowledge in specific fields of specialization and mimic some kind of reasoning process to solve problems in those specialized fields. Each specific field of specialization is called a *domain*. Medicine, economics, sports, education, entertainment, etc. are some examples of domains, each of which may contain several subdomains.

In any domain, there are experts who have a vast amount of specialized knowledge and skill in applying that knowledge to solve problems in that domain. An Expert System (ES) or Knowledge-based System (KBS) is a computer software system that stores a vast amount of expert knowledge from a domain and exercises reasoning to draw conclusions from a given set of inputs. This chapter introduces the reader to the detailed process of designing and constructing an Expert System. It also introduces the PROLOG programming language which is widely used to code Expert Systems.

3.2 The knowledge pyramid

Fig. 3.1 shows a knowledge pyramid. The lowest category in the pyramid is called *noise*. It is a collection of useless and meaningless bits of information that comes hand in hand with data, and is undesirable because it distorts the original data. *Data* refer to the raw figures without any units attached to them. For example, "20" is just a piece of data. One cannot infer anything significant from this piece of data, other than the fact that it represents some measurable or countable quantity. When we attach some units to the raw figures they become *information*. For example, 10 points, 15 cm, 20 years or 100 individuals are pieces of information. The distinction between data and information is often blurred, though. What is commonly referred to as a data-base may be called an information-base and what is commonly referred to as information processing may as well be called data processing.

Associating information with specific objects becomes *knowledge*. It is knowledge that makes an action possible. "I need a 15 cm LAN cable;" "she has been manager for 16 years;" "in Japan you officially become an adult at the age of 20" are typical examples of what we mean by knowledge. Based on knowledge, humans make inferences and draw conclusions. For example, given the above bits of knowledge, one may draw inferences, like: "Since my wired-router is very close to my PC, a 15 cm LAN cable would be sufficient;" "given the fact that she has been a manager for 16 years, she is quite experienced in business;" "if you are an adult you are legally permitted to consume alcohol, although we recommend you refrain from it as much as possible." The highest level of knowledge is *wisdom*. It is qualitatively different from all other types of knowledge. In human society, wisdom is the source of moral character and right action.

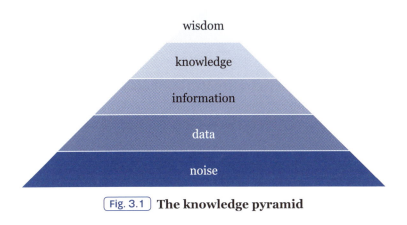

Fig. 3.1 **The knowledge pyramid**

In summary, *noise* is meaningless figures; *data* are raw figures; *information* is meaningful figures; *knowledge* is useful information and *wisdom* is transcendental knowledge. Note that this a somewhat over-simplified way of neatly classifying data, information and knowledge. The distinct categories could overlap depending on the context and usage.

Table 3.1 shows how humans and computers handle noise, data, information, knowledge, and wisdom. Humans ignore noise, while computers filter it. Humans interpret data, while computers store them in a database, analyze and retrieve them. Humans attach some meaning to information, while computers use information to build Information Systems. For humans, knowledge is useful for action, while in computers it is used to create Expert Systems. Up to the knowledge level in the pyramid, computers rival humans. However, will future AI programs be capable of wisdom? This is the question all AI researchers must face as they ponder over the ethical ramifications of their research.

Table 3.1 **Knowledge as handled by humans and computers**

Knowledge pyramid	As handled by humans	As handled by computers
Noise	ignore	filter
Data	interpret	store, analyze, and retrieve
Information	attach meaning	Information Systems
Knowledge	invoke action	Expert Systems
Wisdom	take moral decisions	future possibility?

3.3 Building Expert Systems

Three groups of people are associated with the development, and use deployment of an Expert System.

(1) The team of experts responsible for providing the domain knowledge necessary for constructing the knowledge base.

(2) The team of knowledge engineers responsible for eliciting

knowledge from the experts, for representing the knowledge in a formal way, and for constructing a suitable inference engine.

(3) The team of end users consulting the Expert System to get their problems solved.

The three major steps in building an Expert System are: Knowledge acquisition, knowledge representation and inference (Fig. 3.2). These are described in the following sub-sections.

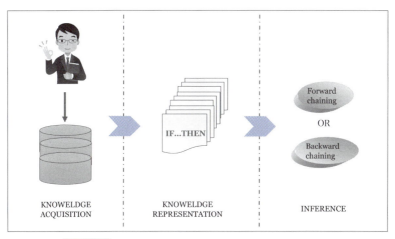

Fig. 3.2 **Steps in building an Expert System**

3.3.1 Knowledge acquisition

This is the very first step in building an Expert System. The knowledge engineer meets with the domain expert to gather the knowledge essential for the construction of the knowledge base. The procedures for knowledge gathering are varied. It could be in the form of an interview, filling out a questionnaire, listening to a talk or presentation made by the expert, etc. In practice, the knowledge engineer invariably encounters a problem known as the knowledge acquisition bottleneck. It is hard to acquire the knowledge as planned. The cause of the bottleneck is two-fold: On the part of the expert, it may be difficulty in articulating the expert knowledge and/or reluctance to part with it, while on the part of the knowledge engineer, it may be difficulty in comprehending, formalizing and representing the expert knowledge.

3.3.2 Knowledge representation

The knowledge acquired from the experts may be in the form of unstructured text and diagrams. This knowledge is then represented in various formats to remove ambiguities and make the coding work easier. Semantic networks and frames are the two well-known knowledge representation formats described below.

Semantic network

Semantic networks use a graph of labeled nodes and labeled directed arcs to encode knowledge. The nodes represent objects, concepts, or events, while the arcs represent the relationships between the nodes. The node-arc-node relationship is often of the form S-V-O. For example, the semantic network in Fig. 3.3 gives a visual representation of the following knowledge snippet: Tom and Sally are classmates, majoring in French. Although Tom hates all other kinds of sports, he likes cycling, which is a kind of sport. Sally, on the other hand, loves sports and goes cycling sometimes.

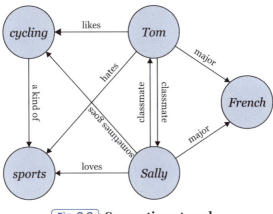

Fig. 3.3 **Semantic network**

Frame

Knowledge representation in the form of frames is more structured than that in semantic networks. A frame contains slots and fillers. Slots hold the attributes and fillers hold the corresponding values of the attributes. The value of a slot can be another frame.

Book Frame	
Title	A Brief History of Time
Author	Stephen Hawking
Publisher	Bantam
Year	1998

Fig. 3.4 **Book frame**

For example, in Fig. 3.4, the book frame contains slots like title, author, publisher and year with the corresponding slot values of 'A

Brief History of Time', 'Stephen Hawking', 'Bantam', and '1998'. Often, the slot values themselves contain other entities. One writes separate children frames for each of these entities and links them to the parent frame. In Fig. 3.5, the student frame contains name, ID, major, money and hobby slots. Slots can be multi-valued as in the case of money in the student frame. The money slot values in the above example imply Taro has money in the form of cash and as deposits in the post-office and bank accounts. The bank account slot value links to the child bank account frame. Similarly, the tennis frame slot links to the child tennis frame containing other slots with their values.

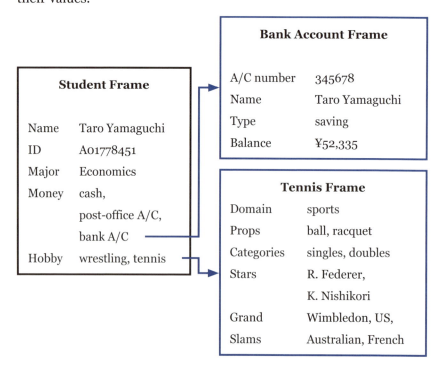

Fig. 3.5 Frames linked through slot values

The visual representation of knowledge as semantic networks or frames is stored in the knowledge base in the form of facts and rules as described below.

Facts

Facts refer to the certainty of objects and their relations in the world. The following are examples of individual knowledge facts:

Tom is fat.

Grass is green.

Mary is the mother of Bill.

Rules

The knowledge rules represent a relation among the knowledge facts. These relations are framed as IF-THEN rules. These rules express simple IF (*antecedent*) - THEN (*consequent*) relationships as in:

IF traffic_light (*green*) THEN (*proceed*).

IF traffic_light (*red*) THEN (*stop*).

IF traffic_light (*yellow*) THEN (*proceed_with_care*).

The rules may also contain conjunctions as shown below:

IF symptom1 (*cough*) AND symptom2 (*running_nose*)
THEN (*cold*).

IF temperature (*high*) OR humidity (*high*)
THEN (*uncomfortable*).

3.4 Inference

The rules and facts in the knowledge base (KB) are static. The user supplies dynamic facts when asked by the Expert System. Inference refers to the process of drawing conclusions by matching the facts provided by the user to the antecedents of the rules in the knowledge base. The rules which satisfy the matching conditions are placed on a queue called the agenda. The agenda is a list of all the rules whose antecedents are currently satisfied. By default, the rules on the agenda are fired in the order in which they happen to be. This firing of the rules produces new facts. The inference engine then searches for rules in the KB which the newly produced facts match. If found, they are placed on the agenda to be fired. The *search-match-fire* cycle of the inference engine continues until no matches are found in the KB.

Just like human experts, Expert Systems do not reach the final conclusion in one step. They follow a chain of linked reasoning to come to a conclusion. First, the given data are matched to the antecedents of the rules. If there is a match, the rule fires, producing more data. These data are then matched to other rules in the KB. They, in turn, fire if matched. This produces a chain of antecedents-consequences, eventually leading to the final conclusion. The process of linking the rules while inferencing in an attempt to draw the final conclusion is called chaining. Chaining is of two types – Forward Chaining (FC) and Backward Chaining (BC).

3.4.1 Forward Chaining

The inference engine tries to find facts that match with the IF part of the rule. When such facts are found, the IF-THEN rule fires and the consequent of the rule becomes an *assertion* or another fact. The inference then uses the additional facts to fire other rules in the knowledge base. This process of chaining the IF part of the rules to draw the final conclusion is called forward chaining. Since forward chaining works by matching facts (data) with the antecedents of the rules, it is also called data-driven (Fig. 3.6).

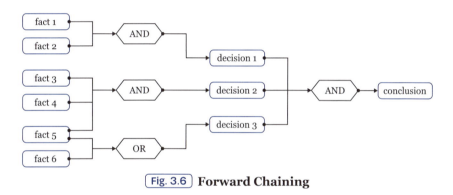

Fig. 3.6 **Forward Chaining**

3.4.2 Backward Chaining

In BC, you hypothesize that the goal is true and then work backwards in finding data that lead to the goal (Fig. 3.7). It is like a tail-to-head search.

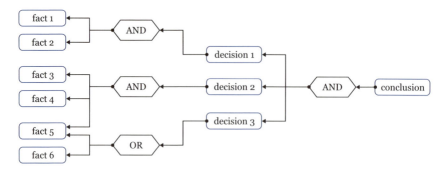

Fig. 3.7 **Backward Chaining**

For a set of given facts, FC and BC result in the same solution. The only difference is the computational cost. As a rule of thumb, one uses FC when there are many output hypotheses and many data up front, and BC when there are fewer output hypotheses and the system must query for data. The branching factor is another important criterion to be considered. Expert Systems are usually designed with an inference engine that goes in the direction with a lower branching factor.

Animal Identification Expert System

This is an example illustrating the FC inference for a system containing rules R1 to R10 in the KB as shown below.

R1: *If* X gives milk *Then* X is mammal.

R2: *If* X has hair *Then* X is mammal.

R3: *If* X is mammal *And* X eats meat *Then* X is carnivore.

R4: *If* X is mammal *And* X has pointed teeth

And X has forward-pointing eyes *Then X* is carnivore.

R5: *If X* is mammal *And X* has hoofs *Then X* is ungulate.

R6: *If X* is mammal *And X* chews cud *Then X* is ungulate.

R7: *If X* is carnivore *And X* has tawny color
And X has black stripes *Then X* is tiger.

R8: *If X* is carnivore *And X* has tawny color
And X has dark spots *Then X* is cheetah.

R9: *If X* is ungulate *And X* has white color
And X has black stripes *Then X* is zebra.

R10: *If X* is ungulate *And X* has tawny color *And X* has dark
spots *And X* has long legs *And X* has long neck *Then X* is
giraffe.

Now imagine a group of elementary school students on a trip to a zoo. They see some rare species of animals and consult the Animal Identification Expert System through its touch screen. In the following interaction between the students (SS) and the Expert System (ES), we can see the Forward Chaining inference at work. The intermediate facts are kept in the working memory (WM).

ES: "Hello and welcome to the Animal Identification Expert System. Kindly answer the questions below and I will help you identify the animal you have just spotted."

ES: "Does the animal give milk to its young ones?" (R1)

SS: "We don't know." (R1 does not fire.)

ES: "Does the animal have hair on its body?" (R2)

SS: "Yes." (R2 fires; the fact that X is mammal is added to the WM.)

ES: "Does the animal eat meat?" (R3)

SS: "Yes." (R3 fires; the fact that X is carnivore is added to the WM.)

ES: "Does the animal have pointed teeth & forward-pointing eyes?" (R4)

SS: "Yes." (R4 fires; the fact that X is carnivore is added to the WM.)

ES: "Does the animal have hoofs?" (R5)

SS: "No." (R5 does not fire.)

ES: "Does the animal chew cud?" (R6)

SS: "No." (R6 does not fire.)

ES: "Does the animal have tawny color and black stripes?" (R7)

SS: "No." (R7 does not fire.)

ES: "Does the animal have tawny color and dark spots?" (R8)

SS: "Yes." (R8 fires; the fact that X is cheetah is added to the WM.)

ES: "Is the animal ungulate?" (R9)

SS: "No." (R9 and R10 do not fire.)

The inference stops since there are no more rules to fire and outputs the following conclusion, which is in the working memory:

ES: "THE ANIMAL YOU HAVE SPOTTED IS A CHEETAH!"

By default, the FC inference follows the rules in the order in which they are arranged in the knowledge base. Often the rules are grouped in convenient rulesets and sophisticated algorithms are used to determine the order of invoking the rulesets. The above example can also be worked out through BC. For example, the ES can assume that the animal spotted by the students is a cheetah and then use the IF-THEN rules to work backwards toward the hypothesis.

3.5 Structure of an Expert System

A fully functioning Expert System has many modules seamlessly working together to give advice to the end-user. Each module is briefly described in this section (Fig. 3.8).

Knowledge base

The knowledge base is a collection of facts and IF-THEN rules obtained from the domain experts. In practical Expert Systems, it is quite large containing tens of thousands of rules. The rules are arranged into convenient rulesets.

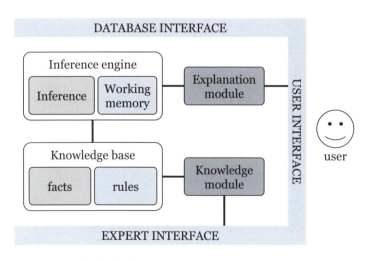

Fig. 3.8 **Expert System structure**

Inference engine

This module is the control unit of the Expert System. It consists of the inference mechanism and a working memory. Facts obtained from the user through query are stored in the working memory. These facts are then matched against the antecedents (IF parts) of the rules and the intermediate conclusions are stored in the working memory. These conclusions become the facts in the next cycle of rule firing. The inference engine could be FC, BC or a mixture of both. The engine keeps running as long as there are rules to fire.

Knowledge module

Knowledge in any field or domain is constantly progressing. The Expert System needs to be regularly updated with new knowledge. This new knowledge is temporarily stored in the knowledge module during the update session. When checked and verified, it is finally transferred to the knowledge base.

Explanation module

Often, the user is not satisfied with the final solution presented by the Expert System. The user may demand an explanation for the conclusions drawn by the Expert System based on the data supplied during the consultation session. The explanation module holds the sequence of the inference steps followed by the Expert System. These may be demonstrated to the user on demand.

User interface

The user interacts with the Expert System via the user interface. It may be the traditional keyboard and mouse interface or an ultra-modern interface with touch screen or voice input.

Expert interface

Experts are not the end users of the system. They normally access the system to update knowledge. Their access point is through the expert interface.

Database interface

The Expert System contains all the basic knowledge and mechanisms to function as an independent and stand-alone application. However, at times it may need to connect to other external databases depending on the queries of the user. This access is through the external database interface.

3.6 PROLOG

PROgramming in LOGic (PROLOG) is a knowledge-manipulating language widely used in programming Expert Systems. It is made up of facts and rules. A fact represents 'what is' and a rule represents a relation. PROLOG predicates and their arguments are in lower-case (Table 3.2). Arguments which begin with capital letters are variables. For example, likes (harry, movies) is a predicate representing the fact that Harry likes movies. On the other hand, likes (Who, What) is a query using the Who and What (or X and Y) variables. The answer to this query would be: Who = harry, What= movies. Likewise, the rules (Table 3.3) always contain variables since they represent generalizations and are instantiated only with the facts provided by the end user querying the Expert System.

Table 3.2 **Knowledge facts**

#	Facts	Meaning
1	male(tom).	Tom is male.
2	female(hanako).	Hanako is female.
3	eats(tom, chocolate).	Tom eats chocolate.
4	likes(hanako, flowers).	Hanako likes flowers.
5	flies(airplane).	Airplane flies.

Table 3.3 **Knowledge rules**

#	Rules	Meaning
1	h(X,Y) :- g(X,Z).	IF g(X, Z) THEN h(X,Y).
2	h(X,Y) :- g1(X,Z), g2(X,Z).	IF g1(X,Z) AND g2(X,Z) THEN h(X,Y).
3	h(X,Y) :- g1(X,Z). h(X,Y) :- g2(X,Z).	IF g1(X,Z) OR g2(X,Z) THEN h(X,Y).

Rule conjunctions

Two antecedents are joined together in a rule using a comma which signifies the AND conjunction.

IF a person is rich AND famous THEN the person is happy.

happy(Person):- rich(Person), famous(Person).

Rule disjunctions

There is no symbol to express OR in PROLOG. A set of rules with identical consequents, but different antecedents serve the function of the OR disjunction as explained below.

IF a person is healthy OR wealthy THEN the person is happy.

happy(Person) :- healthy(Person).

happy(Person) :- wealthy(Person).

3.7 Exercises

1. Give examples of data, information, knowledge and wisdom.

2. Will future computers be wise enough to make moral decisions?

3. Draw a semantic net in any domain of your liking, containing at least ten different nodes. Explain the knowledge contents.

4. Think of your hobbies and other pastimes. Represent them in the form of linked frames.

5. What procedure will you propose to mitigate the knowledge acquisition bottleneck?

6. Why is the explanation module important in an Expert System?

7. Work out the animal identification inference using Backward Chaining inference.

8. Imagine you receive a desperate phone call from your friend saying his laptop is not working. Can you design an Expert System to advise your friend how to troubleshoot the problem?

9. Design an Expert System to suggest a tourist destination depending on the end-user's budget and interests.

10. Write PROLOG facts and rules to reply to the queries of family relations in the family tree shown below (Fig. 3.9).

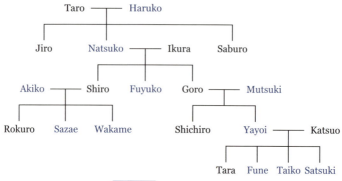

Fig. 3.9 **Family tree**

References

Akerkar, R., & Sajja, P. (2010). *Knowledge-based systems.* Sudbury, MA: Jones and Bartlett.

Clocksin, W. F., & Mellish, C. S. (2003). *Programming in Prolog: Using the ISO Standard.* Springer.

Davies, J-M., Krivine, J-P., & Simmons, R. (Eds.). (2012). *Second Generation Expert Systems.* Springer.

Gelfond, M., & Kahl, Y. (2014). *Knowledge Representation, Reasoning, and the Design of Intelligent Agents: The Answer-Set Programming Approach* (1st ed.). Cambridge University Press.

Giarratano, J. (2004). *Expert systems: Principles and programming* (4th ed.). TBS.

Jackson, P. (1998). *Introduction to Expert Systems* (3rd ed.). Addison-Wesley.

Mohan, C. K. (2000). *Frontiers of expert systems: Reasoning with limited knowledge.* Boston: Kluwer Academic.

CHAPTER FOUR
FUZZY LOGIC AND FUZZY SYSTEMS

4.1 Introduction

In our day-to-day lives, we use expressions like very cold, somewhat cold, quite cold, not very cold to describe the weather and air temperature, for example. What is the *precise* meaning of these different expressions? For instance, what is the real temperature difference between the two expressions *somewhat cold* and *quite cold*? People, when asked to describe the above words using precise numerical values, will be at a loss. This is because our words are imprecise and often ambiguous.

Fuzzy logic is an attempt to model imprecise, ambiguous and vague information. In classical logic, a statement is either completely false or completely true. There is no other possibility in between. However, using fuzzy sets, we talk about the *degree of truth* which can range from 0 (completely false) to 1 (completely true). Therefore, a statement may be true to some degree and false to some degree, all at the same time.

In this chapter, we shall learn about fuzzy sets and linguistic variables. We shall learn how to model fuzzy information with fuzzy membership functions. Based on membership functions, we shall also learn to construct fuzzy inference systems that have wide applications in practical devices and home appliances.

4.2 Non-fuzzy and fuzzy sets

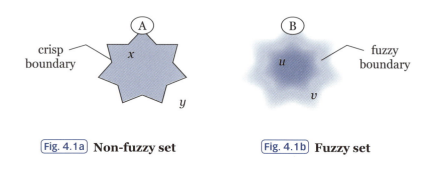

Fig. 4.1a **Non-fuzzy set**   Fig. 4.1b **Fuzzy set**

In a conventional non-fuzzy set (Fig. 4.1a), the boundary is crisp, i.e. sharply defined. An element either belongs to the set (inside or on the crisp boundary) or does not belong to the set (outside the crisp boundary). For example, x belongs to set A, while y does not belong to set A. In other words, membership of x in set A is 1, while that of y is 0. In the fuzzy set B, however, things are quite different. Set B does not have a clearly defined boundary, but rather a gradually fading or fuzzy boundary (Fig. 4.1b). The element u belongs entirely to set B, while the element v belongs only partially. The membership value of u in B is 1, while that of v is in between 0 and 1.

Let us consider a familiar example. In Japan, a person becomes an adult at the age of 20. We can define the term *adult* as a function of the measurable parameter, *age*. If we draw a graph of the function *adult* against *age*, we will get a graph shown in Fig. 4.2. The person under consideration is not an adult at any point that corresponds to the age less than 20. However, exactly at 20, the graph sharply rises from 0 to 1. The equations for adult and non-adult will be:

$$f(x) = 0 \qquad 0 \leq x < 20 \qquad (4.1)$$

$$f(x) = 1 \qquad x \geq 20 \qquad (4.2)$$

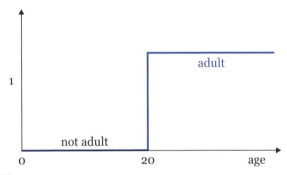

Fig. 4.2 **Non-fuzzy way of defining *not adult* and *adult***

This is the conventional non-fuzzy way of thinking. Depending on age, the person is either an adult (age $\geq$ 20 years) or not an adult (age < 20 years).

Now consider the gray zone shown in Fig. 4.3. It would be more natural to think that a person of age 19.9 is *almost an adult* even though he/she has not yet reached 20. Similarly, it would be perfectly natural to think that a person of age 20.1 is *still not fully an adult*, even though he/she is over 20. The degree of being an *adult* gradually rises, while at the same time, the degree of being a non-adult gradually falls in the gray zone.

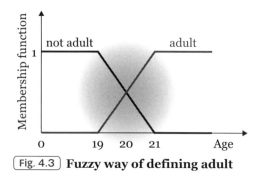

Fig. 4.3 **Fuzzy way of defining adult**

Our natural way of thinking as explained above is *fuzzy*. The two contrasting sets, *not adult* and *adult* are fuzzy sets. Their membership functions plotted in Fig. 4.3 are given by:

not adult function:

$$f(x) = 1 \qquad\qquad 0 \le x \le 19 \qquad\qquad (4.3)$$

$$f(x) = \frac{21 - x}{2} \qquad\qquad 19 < x < 21 \qquad\qquad (4.4)$$

$$f(x) = 0 \qquad\qquad x \ge 21 \qquad\qquad (4.5)$$

adult function:

$$f(x) = 0 \qquad\qquad 0 \le x \le 19 \qquad\qquad (4.6)$$

$$f(x) = \frac{x - 19}{2} \qquad\qquad 19 < x < 21 \qquad\qquad (4.7)$$

$$f(x) = 1 \qquad\qquad x \ge 21 \qquad\qquad (4.8)$$

4.3 Fuzzy systems

This section defines linguistic variables and membership functions. With practical examples, it shows how to design and use fuzzy systems.

4.3.1 Linguistic variables

In mathematics, a variable takes numerical values. For example, the expression speed = 90 kmph consists of a variable *speed* whose value is 90. In contrast, a linguistic variable is a variable whose values are not numbers, but words or sentences. For example, when we describe the age of a person as 14, 19, 30, 55, 70 years, we are using age as a numerical variable. However, if we describe the person as *very young, young, not young, not very old, old*, etc., we are using age as a linguistic variable. Some examples of day-to-day linguistic variables are shown in Table 4.1.

Table 4.1 **Linguistic variables and their values**

Linguistic variables	Linguistic values
temperature	cool, cold, hot, very hot
height	very short, short, tall, very tall
sound	soft, pleasant, noisy, blaring
service	excellent, good, bad, poor
exercise	light, medium, heavy

Linguistic variables are a common occurrence in daily parlance. We use words like cold, freezing, very hot, short, somewhat tall, extremely light, somewhat heavy, etc., without quantifying them. Our speech is naturally *fuzzy* most of the time. Fuzzy logic and fuzzy systems attempt to capture our not-so-precise daily language and make a mathematical model out of it. Linguistic variables, in particular, help us in modeling fuzzy systems as described in the following sub-sections.

4.3.2 Membership functions

Linguistic variables are quantified as membership functions. Consider the linguistic variable age. We may think of its several values like *young*, *middle aged* and *old*. The measurable physical parameter, age is plotted on the x-axis, while the values of the membership functions are plotted on the y-axis. The y-axis values are necessarily between 0 and 1 (Fig. 4.4).

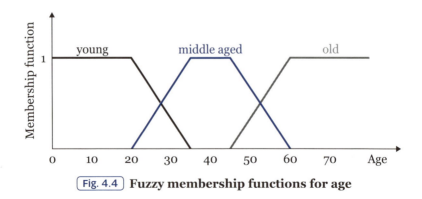

Fig. 4.4 **Fuzzy membership functions for age**

young membership function:

$$f(x) = 1 \qquad\qquad 0 \leq x \leq 20 \qquad\qquad (4.9)$$

$$f(x) = \frac{35 - x}{15} \qquad\qquad 20 < x < 35 \qquad\qquad (4.10)$$

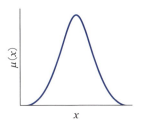
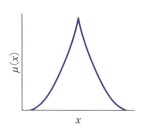
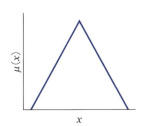
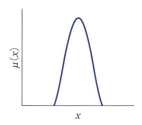

Fig. 4.5 **Various forms of membership functions**

middle-aged membership function:

$$f(x) = \frac{x - 20}{15} \qquad\qquad 20 \leq x < 35 \qquad\qquad (4.11)$$

$$f(x) = 1 \qquad\qquad 35 \leq x < 45 \qquad\qquad (4.12)$$

$$f(x) = \frac{60 - x}{15} \qquad\qquad 45 \leq x < 60 \qquad\qquad (4.13)$$

old membership function:

$$f(x) = \frac{x - 45}{15} \qquad\qquad 45 \le x < 60 \qquad\qquad (4.14)$$

$$f(x) = 1 \qquad\qquad x \ge 60 \qquad\qquad (4.15)$$

In the above example, we have used trapezoid membership functions. Some other types of fuzzy membership functions commonly used to model fuzzy systems are shown in Fig. 4.5.

4.3.3 Fuzzy operations

Just as in the case of the classical sets, fuzzy sets, too, have clearly defined operations (Fig. 4. 6) as:

$$OR \qquad \mu_{A \cup B}(x) = \max\{\,\mu_A(x), \mu_B(x)\,\} \qquad\qquad (4.16)$$

$$AND \qquad \mu_{A \cap B}(x) = \min\{\,\mu_A(x), \mu_B(x)\,\} \qquad\qquad (4.17)$$

$$NOT \qquad \overline{B}(x) = 1 - \mu_B(x) \qquad\qquad (4.18)$$

Following are some examples of fuzzy operations:

- A = {0.9, 0.45, 0.65}
- B = {0.1, 0.35, 0.25}
- A∪B = {max(0.9, 0.1), max(0.45, 0.35), max(0.65, 0.25)}
  = {0.9, 0.45, 0.65}
- A∩B = {min(0.9, 0.1), min(0.45, 0.35), min(0.65, 0.25)}
  = {0.1, 0.35, 0.25}

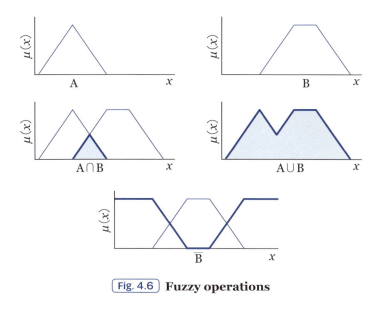

Fig. 4.6 **Fuzzy operations**

4.4 Fuzzy Inference Systems

The design and functional process of a fuzzy controller is shown in Fig. 4.7. The sensor inputs are first converted from analog into digital and then fuzzified by using the relevant fuzzy membership functions. The inference engine fires the relevant fuzzy rules, the output of which is then defuzzied into a crisp value. Finally, the result is converted back to analog at the output. There are two kinds of fuzzy inferences: Mamdami and Takagi-Sugeno. These two are described at the end of this sub-section.

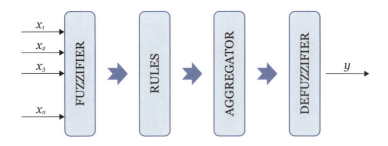

Fig. 4.7 Components of a fuzzy inference system

The steps followed by a fuzzy inference system are:

(1) Define the membership functions of the linguistic variables.

(2) Fuzzify the inputs referring to the input membership functions.

(3) Using fuzzy rules combine the fuzzified inputs to establish a rule strength.

(4) Combine the rule strength and the output membership function to determine the consequence of the rule.

(5) Aggregate the consequences to get an output distribution.

(6) Defuzzify the output using the center of gravity method.

Let us explain each of the components of a fuzzy inference system by working out a practical example. Imagine there exists a system with two inputs, x and y and the output is z. We want to compute the output of the system for some input values of x and y, given the following fuzzy rules:

Rule 1: IF x is A_1 OR y is B_1 THEN z is C_1.

Rule 2: IF x is A_2 AND y is B_2 THEN z is C_2.

Rule 3: IF x is A_3 THEN z is C_3.

Let us consider a practical healthcare example in which the inputs are:

Input x: obesity (measured by BMI)

Input y: cigarette smoking (measured by #packs/day)

Output z: health risk

The fuzzy rules framed by health experts are:

Rule 1: IF obesity is mild OR cigarette smoking is habitual THEN risk is low.

Rule 2: IF obesity is moderate AND cigarette smoking is heavy THEN risk is medium.

Rule 3: IF obesity is extreme THEN risk is high.

The problem is to estimate the health risk for a person whose BMI is x and who smokes y packets of cigarettes a day.

BMI (Body Mass Index) is used to assess obesity. It is computed by dividing body weight measured in kilograms by the square of height measured in meters. Follow the list of steps given below to solve the problem.

<u>Step 1</u>: Defining the membership functions

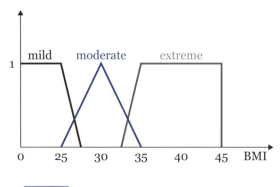

Fig. 4.8 Obesity membership functions

The membership functions expressing obesity are (Fig. 4.8):

Mild obesity: $25 \leq BMI < 30$

Moderate obesity: $30 \leq BMI < 35$

Extreme obesity: $BMI \geq 35$

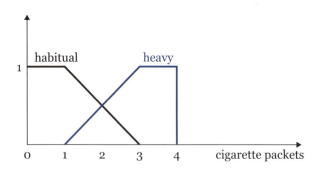

Fig. 4.9 Smoker category membership functions

Smokers may be "habitual" or "heavy," depending on the number of cigarette packets they consume per day. The graphs of the membership functions are shown in Fig. 4.9.

Finally, the health risk associated with obesity and smoking is indicated on a scale ranging from 0 to 100. The membership functions of low, medium, and high risks are shown in Fig. 4.10.

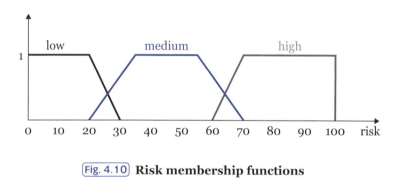

(Fig. 4.10) **Risk membership functions**

Step 2: Fuzzifying the individual inputs (Fig. 4.11)

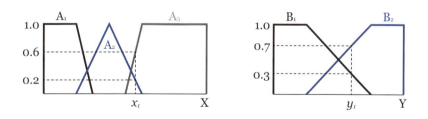

(Fig. 4.11) **Fuzzification of the inputs x and y**

Step 3: Combining the fuzzified inputs (Fig. 4.12)

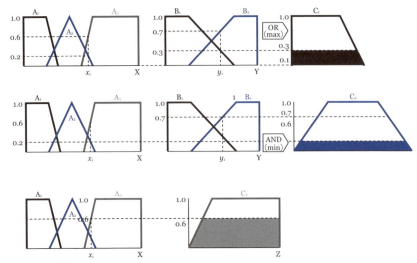

Combining the fuzzy inputs to obtain rule strength

Rule 1: IF x is A_1(0.0) OR y is B_1(0.7) THEN z is C_1(?)

Rule 2: IF x is A_2(0.2) AND y is B_2(0.7) THEN z is C_2(?)

Rule 3: IF x is A_3(0.6) THEN z is C_3(?)

Step 4: Aggregating the consequences (Fig. 4.13)

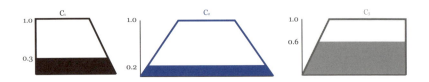

Fig. 4.13 **Aggregating the consequences of the rules**

Step 5: Getting output distribution (Fig. 4.14)

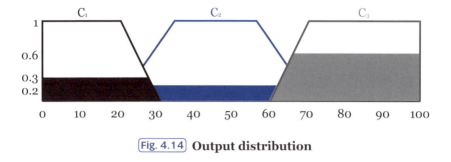

Fig. 4.14 **Output distribution**

Step 6: Defuzzification by calculating center of gravity (COG)

$$COG = \frac{\int_a^b \mu_A(x)x\,dx}{\int_a^b \mu_A(x)\,dx} \qquad (4.19)$$

$$= \frac{(0+10+20)\times0.3+(30+40+50+60)\times0.2+(70+80+90+100)\times0.6}{0.3+0.3+0.3+0.2+0.2+0.2+0.2+0.6+0.6+0.6+0.6}$$

$$= 60.73$$

Mamdani inference

The fuzzy IF-THEN rules have an antecedent (IF) and a consequent (THEN). In the Mamdani inference, the linguistic variables in the antecedent as well as in the consequent are fuzzy. The fuzzified inputs are combined according to the fuzzy rules to establish the rule strength. The consequence of the rule is then found by combining the rule strength and the output membership function. The consequences are combined to get an output distribution. Finally, the output distribution is defuzzified. The above example uses Mamdani inference.

Takagi-Sugeno inference

The Takagi-Sugeno model is an approximation of the Mamdani controller. The linguistic variables in the antecedent part of the rules are fuzzy; however, the linguistic variables in the consequent part of the rules are not fuzzy. Instead, the consequent part of the rules is replaced by a function. This makes the computation rapid, especially in higher dimensional problems.

For the first-order Takagi-Sugeno model, the function is linear.

$$\text{IF } x \text{ is } A \text{ AND } y \text{ is } B \text{ THEN } z = f(x,y) \qquad (4.20)$$

where A and B are fuzzy sets in the antecedent, and $z = f(x,y)$ is a crisp function in the consequent, given by:

$$f(x,y) = ax + by + c \qquad (4.21)$$

For the zero-order Takagi-Sugeno model, the function is constant.

$$\text{IF } x \text{ is } A \text{ AND } y \text{ is } B \text{ THEN } z = k \qquad (4.22)$$

where k is a constant.

In this case, the output of each fuzzy rule is constant. All consequent membership functions are represented by singleton spikes (Fig. 4.15).

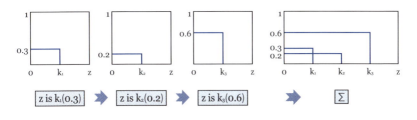

Fig. 4.15 **Takagi-Sugeno inference**

The final (crisp) value is the weighted average given by:

$$WA = \frac{\mu_1(k_1) \times k_1 + \mu_2(k_2) \times k_2 + \mu_3(k_3) \times k_3}{\mu_1(k_1) + \mu_2(k_2) + \mu_3(k_3)} \qquad (4.23)$$

The final risk in the previous example following the Takagi-Sugeno inference will be:

$$WA = \frac{0.3 \times 10 + 0.2 \times 45 + 0.6 \times 85}{0.3 + 0.2 + 0.6}$$

$$= 57.27$$

In summary, the Takagi-Sugeno model is an approximation of the Mamdani inference. It accounts for the fuzziness of linguistic variables only in the antecedents of the rules, but ignores them in the consequent, whereas Mamdani inference considers the fuzziness of the variables appearing both in the antecedents as well as the consequent.

4.5 Exercises

1. Explain with examples: "Fuzzy logic is multi-valued."

2. Give several examples of fuzzy systems.

3. Draw the graphs of membership functions of the above examples.

4. Write down the equations of the membership functions in Fig. 4.5.

5. Work out the fuzzy AND, OR, and NOT operations for the pairs of fuzzy membership functions in Fig. 4.5.

6. Prove De Morgan's law for fuzzy systems.

7. We identify speeds of moving objects as being slowest (e.g. slug), slow (e.g. tortoise), fast (e.g. horse) and fastest (e.g. cheetah) (Fig. 4.16). Draw graphs of the membership functions.

| slowest | slow | fast | fastest |

Fig. 4.16 **Relative speeds of animals**

8. Consider an electric cooler that automatically controls its fan speed according to the temperature of the room. Draw the graph of membership functions for {cold, cool, warm, hot, very hot} depending on the temperature of the room as the input parameter. Design appropriate fuzzy rules relating temperature to fan speed. Fuzzify the inputs for some distinct temperatures

and compute the corresponding fan speeds.

9. Consider a fuzzy washing machine controller shown in Fig. 4.17.

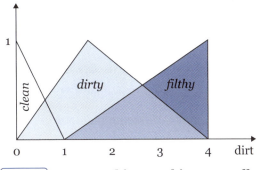

<u>Fig. 4.17</u> **Fuzzy washing machine controller**

The fuzzy rules for the washing time are as follows:

IF clothes (*clean*) THEN washing time = (*10 minutes*).

IF clothes (*dirty*) THEN washing time = (*40 minutes*).

IF clothes (*filthy*) THEN washing time = (*80 minutes*).

Calculate the washing time for the clothes whose dirt reading is:
(a)1.3, (b)2.7, and (c)3.5.

10. It is said that the probability of a university student taking a particular course is directly proportional to the interestingness and weirdness of the professor, but inversely proportional to his strictness. Design a fuzzy inference system and compute the probability of students taking a particular course. Compute your results using Mamdani as well as Takagi-Sugeno inferences.

References

Baldwin, J. F. (1996). *Fuzzy logic*. Chichester, England: Wiley.

Conley, D. (2002). *Fuzzy logic*. Kansas City: Andrews McMeel Pub.

Dimitrov, V., & Korotkich, V. (2002). *Fuzzy logic: A framework for the new millennium*. Heidelberg: Physica-Verlag.

George, J. K., & Yuan, B. (2009). *Fuzzy Sets And Fuzzy Logic: Theory And Applications*. Phi Learning Pvt. Ltd.

Gerla, G. (2001). *Fuzzy logic: Mathematical tools for approximate reasoning*. Dordrecht: Kluwer Academic.

Harris, J. (2000). *An introduction to fuzzy logic applications*. Dordrecht: Kluwer Academic.

Nguyen, H. T., & Walker, E. (1997). *A first course in fuzzy logic*. Boca Raton: CRC Press.

Patyra, M. J., & Mlynek, D. M. (1996). *Fuzzy logic: Implementation and applications*. Chichester: Wiley.

Ross, T. J., Booker, J. M., & Parkinson, W. J. (2002). *Fuzzy logic and probability applications: Bridging the gap*. Philadelphia, PA: Society for Industrial and Applied Mathematics.

Ross, T. J. (2010). *Fuzzy Logic with Engineering Applications* (3rd ed.). Wiley.

Trillas, E., & Eciolaza, L. (2015). *Fuzzy Logic: An Introductory Course for Engineering Students*. Springer.

Zadeh, L. A. (1988). *Fuzzy logic*. Stanford, CA: Stanford University. Center for the Study of Language and Information.

CHAPTER FIVE
WEB MINING AND MACHINE LEARNING

5.1 Introduction

Data Mining is an old statistical technique for finding patterns of information or useful knowledge in large databases. In recent years, the classical statistical techniques like classification, clustering, and association rules have been extended to the largest data repository in the world – the World Wide Web. Web content mining, web structure mining and web usage mining are the three main sub-topics in Web Mining. There is a concentrated effort to develop new Data Mining techniques to cope with the gigantic volume of semi-structured data on the Web.

Machine Learning deals with developing AI programs that learn to extract useful patterns and knowledge from datasets. Supervised learning using Artificial Neural Networks and unsupervised learning using Self Organized Maps are explained in detail in this chapter with several worked out examples. In spite of their phenomenal success, Artificial Neural Nets have been found to be unsuitable for performing Machine Learning in voice and image recognition, because of their shallow structure. Deep Learning with the associated technology of Convolutional Nets is rapidly filling this gap. We are also going to learn about this upcoming field towards the end of this chapter.

5.2 Web Mining

Web Mining is a relatively new research area in AI. It deals with the use of well-established data mining techniques to automatically discover and extract information from Web documents and services. It is an intelligent analysis of the information and services lying across the depth and breadth of the World Wide Web.

The fact that the Web data is massive, semi-structured, heterogeneous, redundant and dynamic in nature makes Web Mining a challenging task. In general, Web Mining tasks are classified into three categories: Web content mining, Web structure mining, and Web usage mining (Fig 5.1).

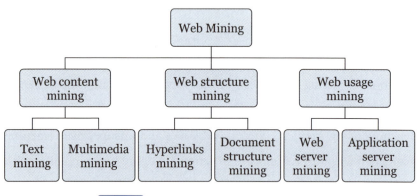

Fig. 5.1 Web Mining classification

5.2.1 Web content mining

Web content mining deals with knowledge discovery, in which the main objects are collections of text documents and, more recently, also collections of multimedia documents. Accordingly, Web content mining can be divided into text mining (including text files, HTML documents, etc.) and multimedia mining (including images, videos, audios, etc., which are embedded in or linked to the Web pages). Text mining is further divided into text categorization (when classes are known in advance) and text clustering (when classes are not known in advance). Multimedia mining deals with the extraction of implicit knowledge, multimedia data relationships as well as patterns not explicitly stored in the multimedia files.

5.2.2 Web structure mining

Web structure mining is the process of discovering structure information from the Web. For this, the Web structure is first represented as a graph. In the Web graph, the Web pages are nodes and the hyperlinks are edges connecting the linked pages. The appropriate handling of the links can indicate potential correlations, and then improve the predictive accuracy of the learned models. *HITS* and *PageRank* are the two conventional algorithms used in Web structure mining. Web structure mining can be further divided into external structure (hyperlinks between web pages) mining, and internal structure (of a web page) mining.

5.2.3 Web usage mining

Web usage mining is the application of Data Mining techniques to discover meaningful patterns in the way users interact with Web pages. The usage mining techniques can predict the behavior of users browsing through the Web. Web usage mining collects the data from Web log records to discover the users' access patterns of Web pages. From the Web server logs, one can easily learn which users access which pages of a Web site during a specified period of time. Depending on the usage data, Web usage mining itself can be further classified into Web server Data Mining (user logs collected by the Web server), and Application server Data Mining (containing business events and application server logs).

Currently, most of the above Web Mining activities are performed in isolation. The "killer-app" devoted to the combined task of Web content, structure, and usage mining is a long way off.

5.3 Machine Learning

Many of our daily computer related activities are made possible by the Machine Learning algorithms that are constantly running behind the scenes as we surf the web, view the photos of our families and friends, and read and respond to emails, just to mention a few. Learning algorithms are used routinely in computer vision, speech recognition, and in natural language processing. In this chapter, we shall learn about Surpervised, Unsupervised, and Deep Learning.

5.3.1 Supervised learning

Supervised learning makes use of the known output from data. The output acts as a supervisor in checking the performance of the program as it is learning. The learning algorithm repeatedly reduces the error between the known output and the one calculated by the program. Learning is achieved when the error becomes minimal.

Table 5.1 **Students' characteristics and their grades**

ID	Attendance	Study	Hobbies	Drinks	Grade
A20170	7	5	5	5	D
A20171	15	10	4	2	A
A20172	12	9	5	0	B
A20173	13	8	6	0	B
A20174	5	4	10	1	F
A20175	9	6	6	4	C
A20176	14	7	5	1	A

The performance of the learned program is tested using test data. It would be ideal to perform the test on new and fresh data. However, in practical situations when new data is not available, the original dataset is divided into a training dataset (roughly 70%) and a testing dataset (roughly 30%). Table 5.1 shows a dataset containing the characteristics of students in a particular class vis-à-vis their grades. The features (attributes) are: classes attended out of a total of 15 per semester, study hours per week, hours devoted to hobbies per week, weekly drinking on a scale of 0-10, and grades from A-F. The AI learning algorithm will automatically extract the information from the feature values in the dataset and use it to predict the grades of new students given their attributes. This is succinctly done by Artificial Neural Nets described below.

Artificial Neural Networks

The individual cells in the biological brains, called neurons, are interconnected to form a large neural network. Each neuron possesses an input and an output filament called a dendrite and an axon, respectively. The dendrites provide input signals to the cell and the axon transmits output signals to other neurons to which it is connected. Signals can be transmitted unaltered or they can be altered by the synapse which is able to increase or decrease the strength of the connection among the neurons. It is the synapse that causes excitation or inhibition of the subsequent neurons.

Artificial Neural Networks (ANN) are an imitation of the structure and function of the natural neural network of the brain. ANN are multi-layered networks consisting of at least three different layers: the input layer, the hidden layer and the output layer (Fig. 5.2). The number of neurons in the input as well as in the output layer is determined by the problem the ANN is learning to solve. However, the number of neurons in the hidden layers is arbitrary.

Each neuron in a given layer is connected to every other neuron in the successive layer. The connection between any two neurons has an associated weight. In essence, the learning carried on by the network is encapsulated in these interconnection weights. Each neuron calculates a weighted sum of the incoming neuron value, transforms it through an activation function, and passes it on as the input to subsequent neurons. The information processing proceeds from the input layer to the output layer via the hidden layer. Hence the name, *Feedforward Network.*

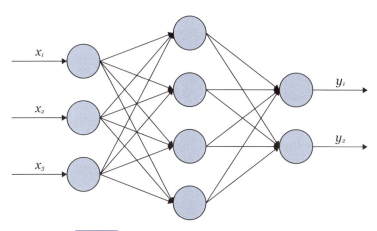

Fig. 5.2 **Feedforward Neural Network**

The n number of inputs in the input layer of the ANN is represented as a vector:

$$X_i = (X_1, X_2, ..., X_n) \tag{5.1}$$

The weights on the connection arcs can be represented by the following weight matrix:

$$W_{ij} = \begin{pmatrix} W_{11} & \cdots & W_{1m} \\ \vdots & \ddots & \vdots \\ W_{n1} & \cdots & W_{nm} \end{pmatrix} \tag{5.2}$$

where, n is the number of neurons in the ith layer and m is the number of neurons in the next j^{th} layer.

The summation of the inputs at the hidden layer neurons is:

$$S_j = \sum_{i=1}^{n} \sum_{j=1}^{m} X_i W_{ij} + b \qquad (5.3)$$

where b is the bias.

The sum of the weights are passed through an activation function leading to non-linear outputs of the hidden layer neurons. In most cases, the Sigmoid function (Fig. 5.3) given below is used as the activation function.

$$y_k = \frac{1}{1 + \exp(-aS_k)} \qquad (5.4)$$

where a is a constant.

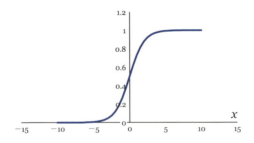

Fig. 5.3 **Sigmoid activation function**

If t_i is the expected output of the ith output neuron and y_i is its actual output, then the total error at the output of the neural net is:

$$Error = \frac{1}{2}(t_i - y_i)^2 \qquad (5.5)$$

The Backpropagation Algorithm for training the neural net, given a training dataset is as stated in Fig.5.4.

Start with randomly chosen weights;
while *Error* is above the desired threshold, do:
    for each input pattern in the training dataset:
        Compute hidden node inputs;
        Compute hidden node outputs;
        Compute inputs to the output nodes;
        Compute the network outputs;
        Compute the error between output and expected output;
        Modify the weights between hidden and output nodes;
        Modify the weights between input and hidden nodes;
    end-for
end-while

Fig. 5.4 **Backpropagation Algorithm**

The working of the Backpropagation Algorithm is illustrated in the training of the following actual neural net (Fig. 5.5). The inputs are two random positive integers and the expected outputs are 1 and 0.

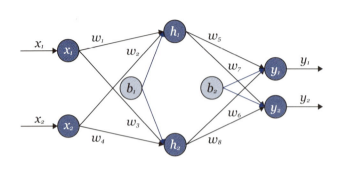

Fig. 5.5 **Backpropagation Algorithm example**

The forward pass

Let us assume we have two inputs, $x_1=2$ and $x_2=5$ and we expect the neural net to give us the outputs $y_1=1$, and $y_2=0$. The inputs summed by the neurons of the hidden layer will be:

$$Sh_1 = w_1 \times x_1 + w_2 \times x_2 + b_1 \times 1$$
$$= 0.4 \times 2 + 0.2 \times 5 + 0.1 \times 1 = 1.9$$

$$Sh_2 = w_3 \times x_1 + w_4 \times x_2 + b_1 \times 1$$
$$= 0.3 \times 2 + 0.5 \times 5 + 0.1 \times 1 = 3.2$$

After applying the activation or the squashing function, we get:

$$h_1 = \frac{1}{1 + e^{-Sh_1}} = \frac{1}{1 + e^{-1.9}} = 0.8699$$

$$h_2 = \frac{1}{1 + e^{-Sh_2}} = \frac{1}{1 + e^{-3.2}} = 0.9608$$

h_1 and h_2 serve as inputs to the neurons in the output layer.

$$Sy_1 = w_5 \times h_1 + w_6 \times h_2 + b_2 \times 1$$
$$= 0.4 \times 0.8699 + 0.1 \times 0.9608 + 0.2 \times 1 = 0.6440$$

$$y_1 = \frac{1}{1 + e^{-Sy_1}} = \frac{1}{1 + e^{-0.6440}} = 0.6557$$

Similarly, $\qquad\qquad y_2 = 0.7871$

Calculating the error in the output layer:

The error at the output y_1 is computed as:

$$Error_{y1} = \frac{1}{2}(t_1 - y_1)^2 = \frac{1}{2}(0 - 0.6557)^2 = 0.2150$$

$$Error_{y2} = \frac{1}{2}(t_2 - y_2)^2 = \frac{1}{2}(1 - 0.7871)^2 = 0.0227$$

$$Error_{total} = Error_{y1} + Error_{y2}$$
$$= 0.2150 + 0.0227 = 0.2377$$

$$\delta_1 = y_1 \times (1 - y_1) \times (t_1 - y_1)$$
$$= 0.6557 \times (1 - 0.6557) \times (0 - 0.6557) = -0.1480$$

$$\delta_2 = y_2 \times (1 - y_2) \times (t_2 - y_2)$$
$$= 0.7871 \times (1 - 0.7871) \times (1 - 0.7871) = 0.0357$$

Correcting the weights of the output layer:

$$w_{jk}^+ = w_{jk} + \eta \delta_k h_j$$
$$w_5^+ = w_5 + \eta \delta_1 h_1$$
$$= 0.4 + 0.7 \times (-\,0.1480) \times 0.8699 = 0.3099$$

$$w_6^+ = w_6 + \eta \delta_1 h_2$$
$$= 0.1 + 0.7 \times (-\,0.1480) \times 0.9608 = 0.0004$$
$$w_7^+ = w_7 + \eta \delta_2 h_1$$
$$= 0.5 + 0.7 \times 0.0357 \times 0.8699 = 0.5217$$
$$w_8^+ = w_8 + \eta \delta_2 h_2$$
$$= 0.7 + 0.7 \times 0.0357 \times 0.9608 = 0.7240$$

Calculating the error in the hidden layer:

The error at the output h_1 is computed as:

$$\mu_j = h_j \times (1 - h_j) \times \sum_k w_{jk}\delta_k \qquad (5.6)$$

$$\mu_1 = h_1 \times (1 - h_1) \times (w_1\delta_1 + w_3\delta_2)$$
$$= 0.8699 \times (1 - 0.8699) \times (0.4 \times (-\,0.1480) + 0.3 \times 0.0357)$$
$$= -\,0.0055$$
$$\mu_2 = h_2 \times (1 - h_2) \times (w_2\delta_1 + w_4\delta_2)$$
$$= 0.9608 \times (1 - 0.9608) \times (0.2 \times (-\,0.1480) + 0.5 \times 0.0357)$$
$$= -\,0.0004$$

Correcting the weights of the input layer:

$$w_{jk}^+ = w_{jk} + \eta \mu_1 x_1$$

$$w_1^+ = w_1 + \eta\mu_1 x_1$$
$$= 0.4 + 0.7 \times (-0.0055) \times 2 = 0.3923$$
$$w_2^+ = w_2 + \eta\mu_1 x_2$$
$$= 0.2 + 0.7 \times (-0.0055) \times 5 = 0.1808$$
$$w_3^+ = w_3 + \eta\mu_2 x_1$$
$$= 0.3 + 0.7 \times (-0.0004) \times 2 = 0.2994$$
$$w_4^+ = w_4 + \eta\mu_2 x_2$$
$$= 0.5 + 0.7 \times (-0.0004) \times 5 = 0.4986$$

The new error given below is an improvement in learning.

$$Error_{total} = 0.2102$$

5.3.2 Unsupervised learning

Unsupervised learning is more difficult, because the program has to learn to cluster the given data without knowing the outcome beforehand. One of the well-known algorithms for clustering is the k-means clustering Data Mining algorithm. The Self-Organizing Map (SOM), described below is also used to achieve clustering.

Self-Organizing Map

The Self-Organizing Map, also called Kohonen network after its inventor belongs to the category of competitive learning networks. When used for clustering data without knowing the class memberships of the input data, the SOM detects features inherent in the data and accordingly forms clusters.

The SOM is a topographic organization in which nearby locations in the map represent inputs with similar properties. SOM maps the multi-dimensional data onto a two-dimensional grid called the map. There are two layers of neurons in SOM: the input layer and the output layer. Each neuron in the input layer is connected to every neuron in the output layer (Fig. 5.6). The SOM is therefore a fully connected network. Further, each output layer neuron has a weight vector associated with it.

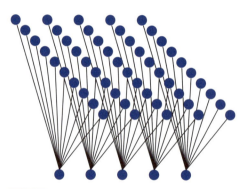

Fig. 5.6 **Self-organizing Map (SOM)**

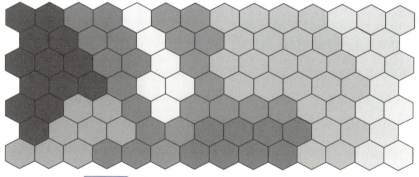

Fig. 5.7 **Clusters visualization using SOM**

The SOM is trained by repeatedly using the input data until stable clusters form. For lack of space, we omit the SOM training algorithm. In the trained SOM, the distance between the adjacent neurons is color-coded as shown in Fig. 5.7. Light areas are clusters and dark areas are cluster separators. This two-dimensional mapping of higher dimensional input data is the outstanding feature of SOM.

5.4 Deep Learning and Convolutional Nets

Regular neural networks are successful in performing classification and prediction over a large number of datasets. However, they are not suitable for recognizing and classifying images. This is because the images are in a 2D format. The 2D format pixel information needs to be converted into a 1D vector form to be fed into the input layer of the regular neural nets. The 2D-1D conversion results in the loss of image information. Secondly, regular neural nets cannot be easily scaled to when extended to image recognition.

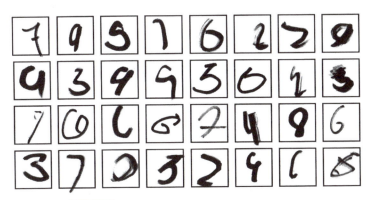

Fig. 5.8 **Handwritten digit recognition**

Convolutional nets are designed specifically for image recognition, such as hand written digits, for example (Fig. 5.8). They exploit the 2D nature of the image pixel information, and since they are not fully connected, the number of parameters (weights, biases, etc.) is far less than those in the regular fully-connected neural nets. The architecture of convolutional nets consists of the following layers:

Input layer

The input layer consists of a 28 x 28 or 32 x 32 square of units, each corresponding to the intensity of a pixel in the input image. However, unlike in the regular neural networks, every input pixel to every hidden is not connected to every other neuron in the next layer. Instead, a small patch of about 4 x 4 neurons in the input layer known as local receptive field or filter is connected to a single neuron in the hidden layer as shown in Fig. 5.9.

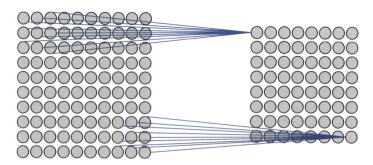

Fig. 5.9 **Input layer units connected to the first hidden layer**

Convolutional layer

The hidden layer following the input layer is called the convolutional layer. The local receptive field in the input layer is slid to the right or down with a convenient stride length, S and is connected to the neighboring neuron in the convolutional layer. When S = 1, we move the local receptive field one pixel at a time; when S = 2, the local receptive field moves by 2 pixels, and so on. Sometimes, it is convenient to pad the input with zeroes around the border. This zero-padding is denoted by P. If W and F are the dimensions of the input layer and the local receptive field, respectively, then the number of neurons that can fit in the convolutional layer is given by:

$$W^* = (W - F + 2P)/S + 1 \qquad (5.7)$$

For example, with W = 28, F = 4, P = 0, and S = 2, $W^* = (28-4+0)/2 + 1 = 13$. This implies that for a 28 x 28 size of the input layer, the corresponding convolution layer will be of size 13 x 13.

Through a scheme known as "weight sharing," we can use the set of 4 x 4 weights and 1 bias on all the connections. Weight sharing implies that all the neurons in the first hidden layer detect exactly the same feature, corresponding to different locations in the input image. In pattern recognition jargon this is referred to as the translation invariance of images. If you move the image of say, a mouse tail in a different location, it is still an image of the mouse tail.

The map from the input layer to the hidden layer is called a feature map. For image recognition, we will need more than one feature map. Hence, a complete convolutional layer consists of several different feature maps as shown in Fig. 5.10.

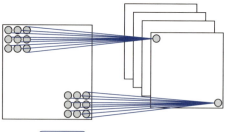

(Fig. 5.10) **Feature maps**

Pooling layer

The convolutional layers are followed by a pooling layer, which takes the output from each feature map in the convolutional layer and downsamples it along the spatial dimensions. For example, each unit in the pooling layer may condense a 2 x 2 region in the preceding layer. The commonly used max-pooling scheme, for example, outputs the maximum activation in the 2 x 2 input region. For each feature map in the convolutional layer there is a corresponding pooling layer.

Fully-connected layer

This layer resembles the output layer of a regular neural net. Each neuron in this layer connects to all the neurons in the previous layer, each connection bearing a unique weight (Fig. 5.11). The Backpropagation Algorithm used in the training of normal artificial neural networks can be used with some minor modifications to train convolutional networks, too.

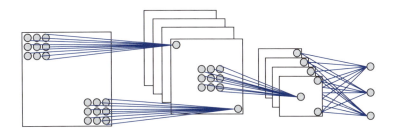

(Fig. 5.11) **CNN architecture**

5.5 Exercises

1. Give examples of Web usage mining that you have come across as a regular Web user.

2. Explain the difference between supervised and unsupervised learning, giving examples.

3. Construct the simplest possible neural net to learn the logic functions AND, OR, XOR. How many layers of neurons will it have? How many neurons will you put in each layer?

4. Starting with random weights, train the above nets by systematically increasing or decreasing the weight values.

5. Use the Backpropagation Algorithm to train the above nets to perform AND, OR, XOR logic functions.

6. What is the role of the bias in Artificial Neural Networks?

7. Construct an ANN for learning to classify the students according to their grades in the dataset of Table 5.1. Starting with random weights, train the ANN using the Backpropagation Algorithm.

What is the prediction accuracy for a new random student?

8. Mention some other applications of SOM.

9. Explain the necessity of the pooling layer in CNN.

10. Mention some other applications of CNN.

References

Alpaydin, E. (2016). *Machine Learning: The New AI*. The MIT Press.

Graupe, D. (2016). *Deep Learning Neural Networks: Design and Case Studies*. WSPC.

Hagan, M. T., Demuth, H. B., Beale, M. H., & De Jesús, O. (2014). *Neural Network Design* (2nd ed.). Martin Hagan.

Han, J., & Kamber, M. (2000). *Data mining: concepts and techniques*. Morgan Kaufmann Publishers.

Hand, D. J., Mannila, H., & Smyth, P. (2001). *Principles of Data Mining*. MIT Press.

Liu, B. (2007). *Web data mining: Exploring hyperlinks, contents and usage data*. Springer.

Loton, T. (2002). *Web Content Mining with Java: Techniques for Exploiting the World Wide Web*. John Wiley & Sons.

Markov, Z., & Larose, D. (2007). *Data mining the web: uncovering patterns in web content, structure, and usage*. Willey-Interscience.

Marsland, S. (2014). *Machine Learning: An Algorithmic Perspective*, (2nd ed.). Chapman and Hall/CRC.

Mitchell, T. M. (1997). *Machine Learning*. New York: McGraw-Hill.

Porter, J. (Ed.). (2016). *Deep Learning: Fundamentals, Methods and Applications*. Nova Science Pub Inc.

CHAPTER SIX
EVOLUTIONARY ALGORITHMS

6.1 Introduction

It is strange that computer scientists often use biological terminolo-gy when speaking of inanimate hardware and software. Take, for in-stance, "computer virus." Is the virus infecting your computer bio-logical? What does it mean to install anti-virus software to kill viruses? What is the meaning of DNA computing? In this chapter, we shall be discussing at length another topic which is central to bi-ology, namely evolution. What do Evolutionary Algorithms have in common with biology?

The terms computer virus, DNA computing, Evolutionary algo-rithms, and so on, are just metaphors. They refer to programs and ways of computing that imitate fundamental biological processes. A living organism infected with a biological virus falls sick and cannot function in a normal way. In the worst case, it dies. A computer in which hackers have stealthily installed malicious programs does not function in a normal way and in the worst case "dies" of hard-disk crash. A computer virus, thus, is an analogy for a malicious software program. DNA computing refers not to the use of living DNA in hardware, but a way of computing that resembles the fundamental processes at work in DNA. Similarly, Evolutionary Algorithms refer to programs that evolve in terms of their performance by artificially imitating the evolutionary process.

6.2 Optimization problems

In chapter 2, we discussed about search problems. In this chapter, we shall have a look at another class of problems, called optimization problems. Given a set of decision variables, how do we determine their values so as to optimize (minimize or maximize) a given objective function?

6.2.1 Optimization problems defined

The first step in solving any problem is to clearly define the problem with all its constraints. An optimization problem may be defined as follows:

$$\text{Find X such that f(X) is minimum/maximum} \qquad (6.1)$$

subject to the constraints:

$$g(X) = 0 \qquad (6.2)$$

$$h(X) < 0 \qquad (6.3)$$

where X is a vector of decision variables, usually bounded as:

$$X_{min} \leq \ X \ \leq X_{max} \qquad (6.4)$$

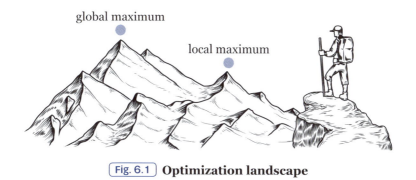

global maximum

local maximum

Fig. 6.1 **Optimization landscape**

f(X) is the objective function, g(X) and h(X) are the constraints, and X_{min} and X_{max} are the bounds on the decision variables in vector X.

The optimization landscape is shown in Fig. 6.1. It is a terrain with flat land, mountains, and valleys. In maximization problems, one seeks to conquer the peaks. Many optimization benchmark functions are designed to trap the optimization algorithms in local maxima. The goal of Evolutionary Algorithms is to find the global maximum within reasonable computational cost. In the minimization problems, one seeks to search for the deepest valley in the search landscape.

6.2.2 Optimization benchmark functions

Optimization involves a great deal of computational complexity. Moreover, many algorithms cannot, in practice, find the global optimum of some functions. Table 6.1 shows a list of minimization benchmark functions designed to test the performance and robustness of optimization algorithms. The Sphere function (#1) is a relatively easy to minimize smooth function, although it becomes difficult with the increasing number of dimensions. The Ackley function (#2) has a deep valley in the middle and several local minima all around, dangerous traps for most hill-climbing (for minimization problems, *valley-descending?*) algorithms. The Rosenbrock function (#3), also known as the valley function, is unimodal. It is relatively easy to find the narrow parabolic valley, but convergence to the global minimum is difficult. The Beale function (#4) is multimodal and has sharp peaks at the corners of the input domain (Fig. 6.2).

Table 6.1 **Optimization benchmark functions**

#	Formulae	Search domain	Min
1	$f(x) = \sum\limits_{i=1}^{n} x_i^2$	$-\infty \leq x_i \leq \infty$ $1 \leq i \leq n$	$f(0,...,0) = 0$
2	$f(x, y) =$ $-20\exp(-0.2\sqrt{0.5(x^2 + y^2)})$ $-\exp(0.5\cos(2\pi x) + \cos$ $(2\pi y))) + e + 20$	$-5 \leq x, y \leq 5$	$f(0, 0) = 0$
3	$f(x) = \sum\limits_{i=1}^{n-1} [100(x_{i+1} - x_i^2)^2$ $+ (x_i - 1)^2]$	$-\infty \leq x_i \leq \infty$ $1 \leq i \leq n$	$f(1, 1) = 0$ $f(1,...,1) = 0$
4	$f(x, y) = (1.5 - x + xy)^2$ $+ (2.25 - x + xy^2)^2$ $+ (2.625 - x + xy^3)^2$	$-4.5 \leq x, y \leq 4.5$	$f(3, 0.5) = 0$

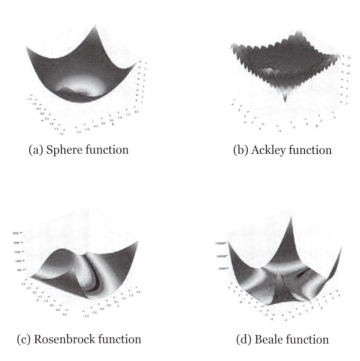

(a) Sphere function   (b) Ackley function

(c) Rosenbrock function   (d) Beale function

Fig. 6.2 **Optimization benchmark functions**

6.3 Evolutionary Algorithms

Evolutionary Algorithms are a successful area of AI. These algorithms are intuitive and easy to program. Their great advantages are that they are robust and domain-independent. This means they depend neither on the problem size nor on the application domain. With minor modifications, they can be used as black boxes to optimize any objective function in any domain, provided the problem can be framed as an optimization problem as illustrated in section 6.2.

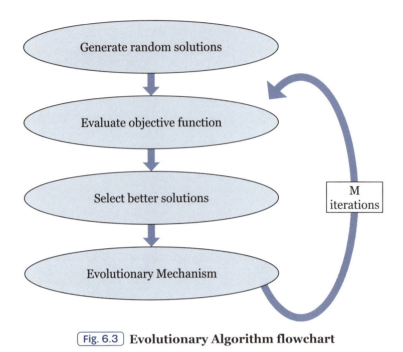

Fig. 6.3 **Evolutionary Algorithm flowchart**

An Evolutionary Algorithm computationally mimics the biological process of evolution. It begins with the random generation of a group of solutions called a population. The individual solutions are then evaluated for their fitness and the better-fit solutions are selected. The fitness which serves as a criterion for survival is directly computed from the objective function. The better-fit solutions are selected and then further evolved using some evolutionary model. Just as in the Darwinian model of natural selection, the lesser-fit solutions are dropped (Fig. 6.3).

6.4 Genetic Algorithm

"Why does the giraffe have such a long neck?" is a question you have probably asked on your first visit to a zoo as a child. Here is a simplified explanation of Darwinian's theory of evolution and natural selection:

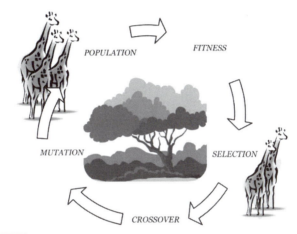

POPULATION   FITNESS

MUTATION   SELECTION

CROSSOVER

Fig. 6.4   **The giraffe's neck evolving over the generations**

In the bad old days, there was a prolonged famine in the African midlands where the ancestors of the present-day giraffe happened to live. Most of the herbivorous animals scrambled for the relatively lower bushes within reach and it was not long before the lower vegetation disappeared, and so did the smaller herbivorous animals. Only the ancestors of the giraffe, who for some reason were slightly taller were "naturally selected" to survive. Over the cycles of generations consisting of selection, crossover of the genetic material and mutations, the best fit, namely the giraffes with the longest necks survived (Fig. 6.4).

We can make a computational model of the above phenomenon, considering the length of the neck of the giraffe as the objective function. The individual traits of the giraffes like age, size, agility, health, etc. and the environmental factors such as temperature, humidity, and availability of water and vegetation could be the relevant input variables for our optimization problem. At first, we will randomly generate a population of giraffes and compute the length of their necks. We would need a functional relationship between the input variables and the length of the neck. According to the evolutionary criterion for survival, the longer the neck, the greater are the chances of survival. Hence, we will define fitness as being directly proportional to the length of the neck. Better-fit individuals will be selected and their attributes (input variables) will be crossed over and mutated to give rise to offspring. The offspring will be tested for fitness and the better-fit individuals will be selected to breed in the next generation cycle, while the lesser-fit individuals will be eliminated. These evolutionary cycles will be repeated for a long time to attain a giraffe with the optimal length of the neck (solution to the optimization problem).

The Genetic Algorithm (GA) precisely follows the evolutionary steps as shown in the flowchart (Fig. 6.5).

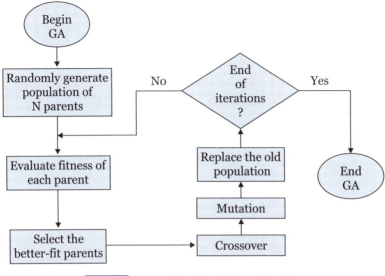

Fig. 6.5 **Genetic Algorithm Flowchart**

Step 1: Random generation of population

A population of N individual solutions is generated randomly. One must ensure that the individuals do not violate the constraints imposed on the input variables. The individual solutions that make up the population are called parents or chromosomes. The biological chromosome is a long strand of bases as shown in Fig. 6.6.

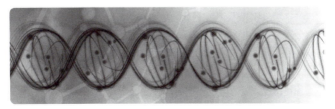

Fig. 6.6 **Biological chromosome**

Each individual solution is a vector X containing the values of the variables. Traditionally, the individuals in the GA population are encoded as binary strings (Fig. 6.7), resembling the DNA structure. Depending on the problem the GA is trying to solve, other encodings are also possible as we shall see later in this chapter.

111000111110001111110001110000110011

111100011101110000110000100001111001

Fig. 6.7 **GA chromosomes encoded as bit strings**

Step 2: Fitness evaluation

The fitness function f(X) is evaluated. In most cases, the evaluation of f(X) is a direct computation. However, in practical application areas, the evaluation of f(X) may involve a time-consuming elaborate simulation. The fitness function for maximization problems is directly proportional to the value of the objective function. For minimization problems, it is customary to consider the reciprocal of the objective function as the fitness.

Step 3: Selection

The greater the value of the fitness function, the fitter the individual. The better-fit parents in the population are selected for crossover (also called recombination). There are two main types of selection schemes: tournament and roulette wheel. The two schemes are described below.

Tournament selection

In the tournament selection, a pair of parents is selected at random from the population. Their fitness is compared and the fitter of the two is selected for "reproduction." In case of a tie in fitness values, the selection is performed randomly. The selection procedure is repeated till the number of the selected parents equals the population size.

Roulette wheel selection

Imagine placing the chromosomes on a roulette wheel according to their fitness and rotating the wheel as in a casino. The chromosomes which occupy the greater area on the wheel have a greater chance of being selected. In practice, this is done as follows (Fig. 6.8).

- Calculate the sum S of all the individual fitnesses.
- Draw a random number r in the interval (0, S).
- Loop through the population and sum individual fitnesses from 0 to s. When s > r, stop the wheel and return the chromosome in that area of the circumference.

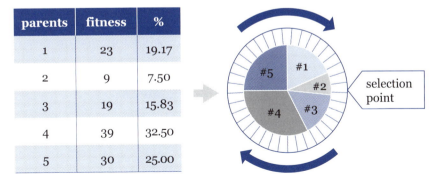

parents	fitness	%
1	23	19.17
2	9	7.50
3	19	15.83
4	39	32.50
5	30	25.00

Fig. 6.8 **Roulette wheel selection**

Step 4: Crossover

In the natural world, crossover mixes the genetic material in the offspring of the species and increases its chances of survival. The following three kinds of crossover operators are common in the GA.

(1) One-point crossover: A single crossover point on both the parents' strings is randomly selected. The part of the chromosome after the crossover point is swapped between the two parent organisms (Fig. 6.9 (a)).

(2) Two-point crossover: Two distinct points are selected on the parent chromosome strings. The part of the chromosome between the two crossover points is swapped between the parent organisms (Fig. 6.9 (b)).

(3) Uniform crossover: Each corresponding bit between the parent chromosomes is swapped with a small probability p (Fig. 6.9 (c)).

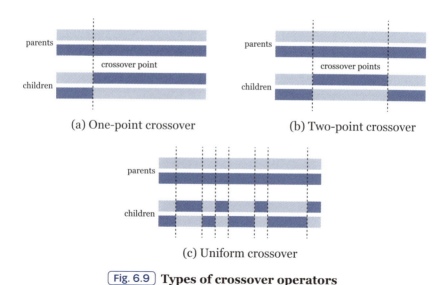

(a) One-point crossover   (b) Two-point crossover

(c) Uniform crossover

Fig. 6.9 **Types of crossover operators**

Step 5: Mutation

Every bit in every individual is modified with a very small mutation probability. Mutation makes the search wider and aids premature convergence of the population.

Step 6: Inserting the new offspring in the older generation

Since we start with randomly generated solutions and evolve them using the GA operators, there is no guarantee that the solutions will always remain within the bounds imposed by the problem constraints. Any infeasible solutions are "repaired" and then inserted back into the old population and the above steps of the GA cycle are iterated.

6.5 GA applications

The GA belongs to the meta-heuristic class of algorithms, which does not depend on the *heuristics* of any particular domain. With some modifications, it can be applied to optimize problems in any domain of interest. Here, we will demonstrate its applicability in solving the Knapsack Problem (KP).

GA solving the Knapsack Problem

Imagine an antique shop where the antique items are displayed, each with its weight and price (Fig. 6.10). Further, imagine a burglar breaking in the antique shop in the middle of the night. He has a small knapsack with limited weight capacity. His aim is to grab as many items as possible that will give him the highest profit. How will he choose the antique items from the displayed collection to fit in the knapsack, so that their total value is maximum? One way would be to start picking up the items that have the largest value. This kind of a *greedy approach* (also known as the *thief's approach*) soon gets stuck in the local maximum. Another reasonable approach would be to make a list of all the possible combinations of items that fit into the knapsack without breaking the capacity con-

1	2	3	4
183 gm	492 gm	333 gm	137 gm
¥ 24,880	¥ 65,430	¥ 43,620	¥ 17,950

5		6
239 gm		70 gm
¥ 32,500		¥ 9,380

7	8	9	10	11
230 gm	87 gm	319 gm	388 gm	348 gm
¥ 30,360	¥ 11,390	¥ 41,470	¥ 52,770	¥ 48,020

Fig. 6.10 **Knapsack problem**

straint and choose the combination that yields the highest value. This approach (also known as the *banker's approach*) is clearly not intelligent. It relies on a brute-force exhaustive search, which may continue forever when choosing items from a large collection.

The GA approach explained below is more elegant. Let us consider the KP shown in Fig. 6.10. There are 11 items, each with a weight and value. Assume the maximum allowable capacity of the knapsack is 1,500 gm. Refer to the steps in the GA flowchart shown in Fig. 6.5.

Step 1: Generation of random population

Let us consider a population of 4 parents (also called *chromosomes*). Each chromosome is encoded as a bit string: bit 1 implies that the item is selected, while bit 0 implies the item is not selected to be put in the knapsack. Each bit constituting the chromosome is called the *gene*. The bits are randomly generated ensuring that the maximum weight capacity of the individual chromosome does not exceed the allowable capacity (Fig. 6.11).

#	1	2	3	4	5	6	7	8	9	10	11	knapsack weight
p1	0	0	1	1	0	0	1	1	1	0	1	1454
p2	0	1	0	1	1	1	0	1	1	0	0	1344
p3	0	0	1	0	0	0	0	0	1	1	1	1388
p4	1	0	1	0	1	0	1	1	0	0	1	1420

[Fig. 6.11] **Population of chromosomes**

Step 2: Evaluation

The KP objective function is the total value of the items put in the knapsack. The value of the objective function for each chromosome is shown in Fig. 6.12. The best fit parent is p1, because it has the highest value.

#	1	2	3	4	5	6	7	8	9	10	11	weight	value
p1	0	0	1	1	0	0	1	1	1	0	1	1454	192810
p2	0	1	0	1	1	1	0	1	1	0	0	1344	178120
p3	0	0	1	0	0	0	0	0	1	1	1	1388	185880
p4	1	0	1	0	1	0	1	1	0	0	1	1420	190770

Fig. 6.12 **Fitness evaluation**

Step 3: Selection

In the tournament selection, a pair of parents from the polulation is picked at random. The fitter of the two is selected for crossover. The selection results are shown in Fig. 6.13. Since p2 is the least fit parent, it is eliminated by the selection procedure.

p3	0	0	1	0	0	0	0	0	1	1	1	1388	185880
p1	0	0	1	1	0	0	1	1	1	0	1	1454	192810
p4	1	0	1	0	1	0	1	1	0	0	1	1420	190770
p3	0	0	1	0	0	0	0	0	1	1	1	1388	185880

Fig. 6.13 **Tournament selection**

Step 4: Crossover

Crossover is performed with a pair of chromosomes at a time. In the uniform operator scheme, each and every bit is swapped between the mating pair with a small probability. The children resulting from the parent chromosomes after crossover are shown in Fig. 6.14.

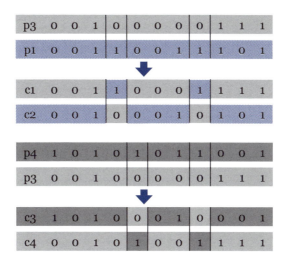

(Fig. 6.14) **Uniform crossover**

Step 5: Mutations

A bit in a chromosome is randomly selected. If it is 1, it is replaced by 0 and vice-versa. This is done for $m\%$ of the genes in the population. While performing mutations, care should be taken to avoid infeasible solutions (i.e. solutions that exceed the knapsack capacity). Fig. 6.15 shows the new population formed after mutating the children. In this new population the best fit parent is p3, which shows a higher KP value than the parents in the initial population.

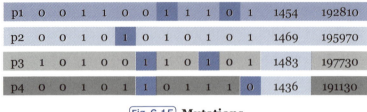

p1	0	0	1	1	0	0	1	1	1	0	1	1454	192810
p2	0	0	1	0	1	0	1	0	1	0	1	1469	195970
p3	1	0	1	0	0	1	1	0	1	0	1	1483	197730
p4	0	0	1	0	1	1	0	1	1	1	0	1436	191130

Fig. 6.15 **Mutations**

The above steps of the GA cycle are iterated many times. The algorithm converges to the optimum solution after several iterations.

6.6 Exercises

1. Explain the role of crossover and mutation operators in the GA.

2. The KP is a classical combinatorial optimization problem. "Knapsack" is just a metaphor used to make the description of the problem interesting. Can you give real-life examples of KP?

3. Optimize the sphere function in Table 6.1 using the GA. Use real number coding.

4. The solution to the N Queens problem is finding an N x N board configuration in which no more than one Queen may be placed on the same row, column or diagonal of the board. Solve the 4 Queens problem using the GA.

5. Solve a *Sudoku* problem using the GA.

6. You are given an N x N checkboard. The objective is to shade every square unit with a color from a set containing at most four different colors, such that no adjacent squares are of the

same color. Solve the problem the using GA (Fig. 6.16).

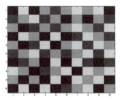

Fig. 6.16 **Checkboard problem**

7. You are given 10 cards numbered from 1 to 10. Use the GA to divide the cards into two piles such that (1) the sum of the numbers on the cards in pile 1 is as close as possible to 36 and (2) the product of the numbers on the cards in pile 2 is as close as possible to 360.

8. With the GA try to create a weekly school timetable, violating the least number of constraints.

9. Tiddly Winks is a popular children's game. Given a board containing a configuration of circles of varying diameter, draw the largest possible non-intersecting circle in the open spaces. Use the GA to solve the Tiddly Winks game problem (Fig. 6.17).

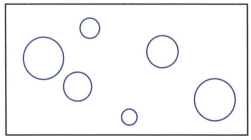

Fig. 6.17 **Tiddly Winks game**

10. How will you train a neural net using the GA?

References

Back, T. (1996). *Evolutionary Algorithms in Theory and Practice: Evolution Strategies, Evolutionary Programming, Genetic Algorithms.* Oxford University Press.

Chambers, L. (2001). *The practical handbook of genetic algorithms: Applications.* Boca Raton, FL: Chapman & Hall/CRC.

Eiben, A. E., & Smith, J. E. (2015). *Introduction to Evolutionary Computing* (2nd ed.). Springer.

Goldberg, D. E. (1989). *Genetic Algorithms in Search, Optimization, and Machine Learning.* Addison-Wesley Professional.

Kumar, A., & Gupta, Y. P. (1995). *Genetic algorithms.* Exeter, England: Pergamon.

Michalewicz, Z. (1994). *Genetic algorithms.* Berlin: Springer-Verlag.

Pham, D. T., & Karaboga, D. (2000). *Intelligent optimisation techniques: Genetic algorithms, tabu search, simulated annealing and neural networks.* London: Springer.

Reeves, C. R., & Rowe, J. E. (2003). *Genetic algorithms: Principles and perspectives: A guide to GA theory.* Boston: Kluwer Academic.

Simon, D. (2013). *Evolutionary Optimization Algorithms.* Wiley.

Sivanandam, S. N., & Deepa, S. N. (2007). *Introduction to genetic algorithms.* Berlin: Springer.

## CHAPTER SEVEN

SWARM INTELLIGENCE

 Introduction

A swarm is a large number of homogenous, unsophisticated agents that interact locally among themselves and with their environment without any central control or management. The collective behavior of self-organized, but decentralized natural or artificial systems, that leads to the solution of complex problems is called Swarm Intelligence (SI). The individuals that make up the swarm are often extremely simple agents, that lack memory, intelligence or even awareness of one another. By following simple rules like sticking together and avoiding collision, they give rise to a form of emergent intelligence. A colony of ants finding the shortest route between the nest and the food source, or a swarm of bees finding spots with the maximum amount of nectar are good examples of SI.

Ant Colony Optimization (ACO) and Particle Swarm Optimization (PSO) are well-known AI optimization algorithms, basing themselves on the SI paradigm. Being meta-heuristic domain independent algorithms, they have become very popular in solving complex optimization problems in various fields. The details of these algorithms along with their applications are discussed in this chapter.

7.2 Ant Colony Optimization

The ACO algorithm is inspired by a colony of ants that start randomly searching for food, and trace the shortest path from the food source to the colony rather quickly. The ant colony behavior is driven by a communication system called *stigmergy* described below.

Fig. 7.1 Random movement of ants

Initially, all the ants in the colony move randomly until some of them find some food (Fig. 7.1).

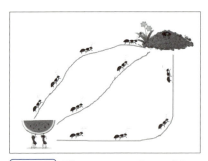

Fig. 7.2 Pheromone deposition on routes

When some ants find food, they carry it back to their nest. On the way back, they lay *pheromone* on the trail they have followed from the food source to the nest. The pheromone laid on the path acts as an indicator to the other ants of the colony (Fig. 7.2).

Fig. 7.3 **Pheromone evaporates**

The ants following the pheromone trails created by their predecessors, in turn, lay more pheromone on the trails. This results in several pheromone-laden trails from the food source to the nest. The pheromone being volatile evaporates with time (Fig. 7.3).

Fig. 7.4 **The shortest path remains**

With the passage of time, the deposits on the longer trails become thinner, while those on the shorter and more convenient paths become thicker as more and more ants deposit pheromone before it has time to evaporate.

Eventually, all the longer paths disappear due to the pheromone evaporation and the shortest path with the thickest layer of pheromone remains to guide the ants from the food source to the colony (Fig. 7.4). Stigmergy (indirect communication through pheromone) enables the ant colony to find the shortest path between the food source and the nest.

The discrete ACO algorithm derived from the above description of the ant colony behavior is shown in Fig. 7.5. It is used to find the shortest route through a given set of nodes.

1. Initialize parameters and solutions
2. While the termination criterion is not met
    3. Evaluate solutions
    4. Update pheromone
    5. Construct new solutions
6. End
7. Output the optimum solution

Fig. 7.5 **Generic ACO algorithm**

In step 1, the algorithm parameters are initialized and all the artificial ants (random solutions) are generated. The loop from steps 2 through 6 is repeated until the termination condition is met. The steps inside the loop consist of evaluating the solutions, updating the pheromones and constructing new solutions from the previous solutions. The two main steps inside the loop are further described below.

Solution construction

Ant k on node i selects node j, based on the probability, p_{ij}, given by:

$$p_{ij}^k = \begin{cases} \dfrac{[\tau_{ij}]^\alpha [\eta_{ij}]^\beta}{\sum_{j \in \mathcal{N}_i^k} [\tau_{ij}]^\alpha [\eta_{ij}]^\beta} & \text{if } j \in \mathcal{N}_i^k, \\ 0 & \text{otherwise.} \end{cases} \tag{7.1}$$

where $\mathcal{N}_i^k$ denotes the set of candidate sub-solutions; τ_{ij} and η_{ij} denote, respectively, the pheromone value and the heuristic value associated with e_{ij}. α and β are constants, showing the relative importance of τ_{ij} and η_{ij}. Heuristic η_{ij} is the reciprocal of the distance d_{ij} between two nodes.

Updating the pheromone

The operator employed for updating the pheromone value of each edge e_{ij} is defined as:

$$\tau_{ij} = (1-\rho)\tau_{ij} + \rho \sum_{k=1}^{m} \Delta\tau_{ij}^k \qquad (7.2)$$

$$\Delta\tau_{ij}^k = \frac{1}{L^k} \qquad (7.3)$$

Where L^k denotes the quality of the solution created by ant k and $\rho \in [0,1)$ denotes the evaporation rate. In practice, L is initially set to a large number.

7.3 Particle Swarm Optimization

The Particle Swarm Optimization (PSO) is based on the food-gathering behavior of a flock of birds or a school of fish or the nectar-gathering behavior of a swarm of bees. Let us imagine there is a swarm of bees trying to find a location with the maximum amount of nectar. Their search space is a large garden as shown in Fig. 7.6.

The behavior of the swarm can be explained as follows:

Initially, the swarm of bees moves randomly trying to find some nectar. The swarm stays together, but fairly dispersed. The bee closest to some flowers finds some nectar (Fig. 7.6).

(Fig. 7.6) **Random movement of the swarm**

The bee who has found the nectar immediately sends a message to the other bees. The rest of the bees then fly towards the signaling bee, but also follow their own instincts (Fig. 7.7).

(Fig. 7.7) **Leader bee signaling**

While flying toward the signal sending bee, and at the same time following its own instincts, some other bee may find another location with a greater amount of honey than the previous bee. This bee then acts as the signal sender (Fig. 7.8).

(Fig. 7.8) **Bees following the leader**

The swarm repeats the above steps many times. In every step, each

Fig. 7.9 **Convergence of the swarm**

bee tries to fly in the direction between the location of the maximum amount of nectar found by the swarm and that of the maximum amount of nectar found by itself. It also flies somewhat randomly.

Following the swarm, and at the same time following its instincts with a bit of randomness, increases each bee's chances of finding a spot with the maximum amount of honey. After repeated trials, the swarm locates the place containing the maximum amount of nectar (Fig. 7.9).

The PSO algorithm conducts a search using a population of individuals. The individual in the population is called a particle and the population is called a swarm. The performance of each particle is measured according to a predefined fitness function. In every iteration, the particles move taking into consideration their previous personal *p-best* as well as the global *g-best*. The process is repeated till the convergence of the swarm.

The notations used in PSO are: the ith particle of the swarm in iteration t is represented by the d-dimensional vector, $x_i(t) = (x_{i1}, x_{i2}, ..., x_{id})$. Each particle also has a position change known as velocity, which for the ith particle in iteration t is $v_i(t) = (v_{i1}, v_{i2}, ..., v_{id})$. In a given iteration t, the velocity and position of each particle is updated using the following equations:

$$v_i(t) = wv_i(t\text{-}1) + c_1 r_1 (pbest_i) - x_i(t\text{-}1)) + c_2 r_2 (gbest - x_i(t\text{-}1)) \qquad (7.4)$$

$$x_i(t) = x_i(t\text{-}1) + v_i(t) \qquad (7.5)$$

where, $i = 1, 2, ..., N$; $t = 1, 2, ..., T$. N is the size of the swarm, and T is the iteration limit; c_1 and c_2 are called social parameters, indicating the degree of confidence in oneself and the swarm, respectively; r_1 and r_2 are random numbers between 0 and 1; w is the inertia weight that controls the impact of the previous history of the velocities on the current velocity, influencing the trade-off between the global and local experiences. A large inertia weight facilitates global exploration, while a small one tends to facilitate local exploitation. Equation 7.4 is used to compute a particle's new velocity, based on its previous velocity and the distances from its current position to its local *p-best* and to the global *g-best* positions. The new velocity is then used to update each particle's new position (Equation 7.5).

7.4 Applications of SI techniques

The above two swarm optimization techniques have found applications in solving complex optimization problems. In this section, we shall illustrate their use in solving a practical problem and optimizing a benchmark function.

7.4.1 ACO optimizing Travelling Salesperson Problem

The Travelling Salesperson Problem[1] (TSP) is a minimal path search problem. Starting from the home city, a salesperson has to go on a tour of N distinct cities before returning to the home city. The salesperson has access to a list of the pairwise distances between each city on the tour. The problem constraints are that the salesperson should visit all the cities without skipping any and that every city should be visited only once. In what order should the salesperson visit the cities so that the tour length is minimal?

The TSP is one of the most cited NP-hard combinatorial optimization problems. It is estimated that it will take several days or even months for a computer to compute all the possible combinations for a relatively small-size TSP. For larger problems, the solution time may be more than the age of the universe!

Many algorithms and techniques have been developed to find the global optimal solution in reasonable time. In this sub-section, we will try to use the ACO meta-heuristic (domain-independent heuristic) algorithm to find the global optimum of a given TSP.

Rather than visiting *cities* on a strenuous business trip, let us consider visiting various spa spots on a leisure trip. Imagine you are on a holiday trip to Japan and you want to visit some of the most famous

---

1 TSP stands for Travelling *Salesman* Problem; the author prefers to rename it as "Travelling *Salesperson* Problem."

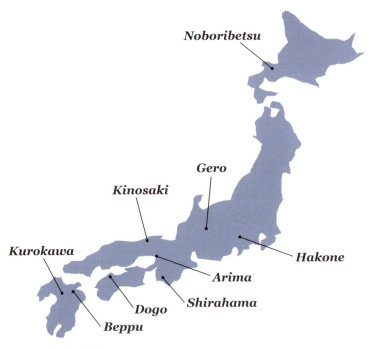

Noboribetsu

Gero

Kinosaki

Kurokawa

Hakone

Arima

Dogo　Shirahama

Beppu

Fig. 7.10 Visiting celebrated hot springs in Japan

Japanese hot springs (*onsen*). Fig. 7.10 shows a map containing 9 celebrated hot springs spread out on the four main islands of Japan. Let's say you have a round-trip air ticket to and from the island of Kyushu in the south. So, your *onsen*-tour begins and ends at the Beppu *onsen*. While you do not mind the expenses of getting yourself immersed in the relaxing waters of an *onsen*, your budget does not allow you a bullet-train ride crisscrossing the country. Therefore, you have to calculate the shortest possible *onsen* round-trip inside the country driving a rented car. The list of your preferential *onsens* is: Arima, Beppu, Dogo, Gero, Hakone, Kinosaki, Kurokawa, Noboribetsu, and Shirahama. The distances between the *onsens* are shown in Table 7.1.

	Arim	Bepp	Dogo	Gero	Hako	Kino	Kuro	Nobo	Shira
Arim	♨	466	303	287	437	131	503	1408	185
Bepp	466	♨	179	890	1041	655	71	1989	629
Dogo	303	179	♨	583	733	362	216	1700	466
Gero	287	890	583	♨	344	360	940	1197	417
Hako	437	1041	733	344	♨	531	1091	1150	554
Kino	131	655	362	360	531	♨	706	1391	327
Kuro	503	71	216	940	1091	706	♨	2038	666
Nobo	1408	1989	1700	1197	1150	1391	2038	♨	1536
Shira	185	629	466	417	554	327	666	1536	♨

Table 7.1 Distances between pairs of *onsens* (km)

Let us find the optimum *onsen* round-trip route using the ACO algorithm. The flowchart is shown in Fig. 7.11.

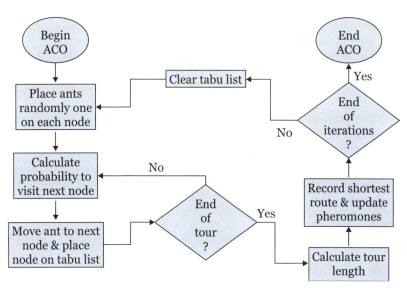

Fig. 7.11 ACO algorithm for solving TSP

Step 1: Since the route consists of 9 nodes, we place 9 artificial ants one on each node. Each of these ants will construct a solution to the problem.

Step 2: Each ant calculates the probability p of visiting the next node from the list of the remaining 8 nodes, by using Equation 7.1.

Step 3: A random number $r \in [0,1)$ is picked and if $r <= p_j$, the ant visits the node j. The selected node j is placed on the tabu list.

Step 4: Each ant repeats steps 2 and 3 till the tour is complete.

Step 5: Each ant calculates the length of the tour it has constructed and the shortest route is recorded.

Step 6: The pheromone on the edges is updated using Equations 7.2 and 7.3.

Step 7: Steps 1 to 6 are repeated for a pre-determined number of iterations.

Table 7.2 shows the optimal solution found by the ACO algorithm. The shortest onsen round-tour is: Beppu(2)-Dogo(3)-Kinosaki (6)-Gero(4)-Noboribetsu(8)-Hakone(5)-Shirahama(9)-Arima(1)-Kurokawa(7)-Beppu(2). The length of the tour is 4,561 km.

(Table7.2) **ACO solutions to the *onsen* problem**

Iter	BEST ROUND TOUR									Dist(km)
0	2	1	4	5	3	8	6	7	9	2	6922
10	2	9	4	8	6	7	3	5	1	2	6192
20	2	1	6	3	4	9	5	8	7	2	5772
30	2	3	6	9	1	4	5	8	7	2	4943
50	2	1	4	8	5	6	9	3	7	2	4711
96	2	3	6	4	8	5	9	1	7	2	4561

7.4.2 PSO optimizing sombrero function

The 2-D sombrero[2] function looks like an elegant Mexican hat (Fig. 7.12). The function is defined as:

$$z = \frac{\sin(\sqrt{x^2 + y^2})}{\sqrt{x^2 + y^2}}, \quad x^2 + y^2 \neq 0 \tag{7.6}$$

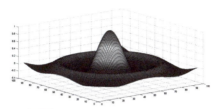

Fig. 7.12 **Sombrero function**

We shall follow the steps outlined in the PSO flowchart (Fig. 7.13) to find the peak of the function.

Step 1: Let us begin by randomly generating a population of 5 particles with their positions in the range [-100, 100] along the x and the y axis. Similarly, let us randomly generate their initial velocities in the range [-10, 10].

Step 2: The objective function of each particle is computed using Equation 7.6. For the maximization problem, fitness is proportional to the value of the objective function.

Step 3: In the zeroth iteration, the personal best (*p-best*) of each particle is the same as its position.

Step 4: The global-best (*g-best*) is the best of the personal-bests

---

2   Sombrero in Spanish means hat.

(*p-bests*).

Step 5: The particle velocities are updated using Equation 7.4. If the velocities exceed the range (-10, 10) they are randomly re-initialized inside the range.

Step 6: The particle positions are updated using Equation 7.5. If the positions exceed the range (-100, 100) they are randomly reinitialized inside the range.

Step 7: Steps 1 to 6 are repeated for a predetermined number of iterations.

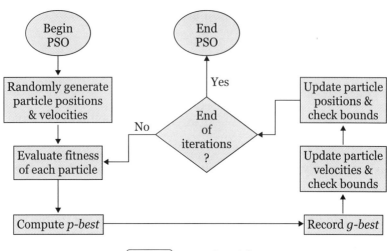

Fig. 7.13 **PSO algorithm**

The convergence of the PSO algorithm is shown in Fig. 7.14. A population of 5 particles with random positons and velocities is generated and their function value is evaluated. These particles are scattered in the search space as shown in Fig. 7.14 (a). In each iteration, the *p-best* and the *g-best* are recorded and the velocities and positions are updated. Fig. 7.14 (b) shows the flying of the particles in the search space after 20 iterations and Fig. 7.14 (c) after 50 iterations. The swarm of particles converges after about 90 iterations (Fig. 7.14 (d)).

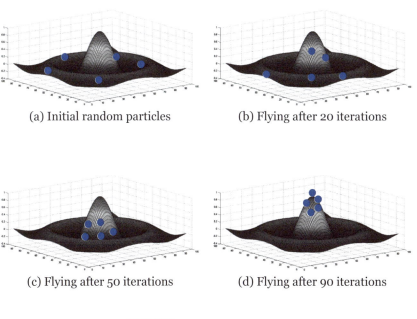

(a) Initial random particles   (b) Flying after 20 iterations

(c) Flying after 50 iterations   (d) Flying after 90 iterations

Fig. 7.14 **PSO convergence**

7.4.3 PSO-ACO hybrid optimizing project management

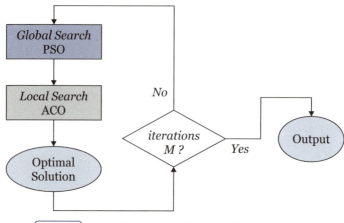

(Fig. 7.15) **PSO-ACO hybrid algorithm flowchart**

Although a continuous ACO and a discrete PSO are both used, the original ACO is discrete and the original PSO is continuous. The classic ACO method of constructing solutions to discrete combinatorial optimization problems piece by piece is not suited to finding solutions to continuous problems. PSO, on the other hand, searches for optimum solutions by flying over continuous spaces. Discretizing PSO results in loss of information due to truncation of the variable values. Using ACO over *continuous* spaces and PSO over *discrete* spaces is stretching the SI metaphor too far. A hybrid algorithm that judiciously combines the strong points of *discrete* ACO with those of the *continuous* PSO would be a classic SI hybrid algorithm (Fig. 7.15). One example is the optimization of a project development schedule, which consists of scheduling various tasks (discrete) and allocating variable time intervals (continuous) to the processing of tasks. The scheduling part is optimized by ACO and the time allotment by PSO as shown in Fig. 7.16.

PROJECT DEVELOPMENT SCHEDULE				
Projects & Tasks	April	May	June	July
Project I Task A				
Project II Task A				
Project I Task B				
Project III Task A				
Project II Task B				
Project III Task B				
Project I Task C				
Project II Task C				
Project III Task C				
Project III Task D				

(Fig. 7.16) **Project management schedule**

7.5 Exercises

1. Like the KP, the TSP is a metaphorical way of describing a problem. Give real-life examples of TSP.

2. Calculate the total number of distinct routes for an N-node TSP.

3. Estimate the computing time for a 100-node TSP using the latest i7 (CPU speed 3.5 GHz) desktop computer.

4. Towers of Hanoi is another NP hard problem. It is easily solved

using *recursion*. Estimate the computing time to solve a 100 disc Towers of Hanoi problem, using the latest i7 desktop computer (Fig. 7.17).

Fig. 7.17 **Towers of Hanoi**

5. Solve the bin-packing problem using ACO.
6. Optimize the benchmark functions in Table 6.1 using PSO.
7. Solve the Tiddly Winks game using PSO (Fig. 6.17).
8. Use the PSO-ACO hybrid to minimize the project development schedule by matching the workers' skills to the tasks.
9. An n x n magic square contains an integer between 1-n in each of the inner square units. Choose the integers in such a way their sums along each row, column, and diagonal are identical. Solve the 3 x 3 magic square problem using ACO (Fig. 7.18).

15	15	15	15	15
15	4	9	2	15
15	3	5	7	15
15	8	1	6	15
15	15	15	15	15

Fig. 7.18 **Magic square (3 x 3)**

10. Solve the same 3 x 3 magic square puzzle using PSO.

References

Blum, C., & Merkle, D. (Eds.) (2008). *Swarm Intelligence: Introduction and Applications*. Berlin: Springer-Verlag.

Bonabeau, E., Dorigo, M., & Theraulaz, G. (1999). *Swarm Intelligence: From Natural to Artificial System*. New York: Oxford University Press.

Bonabeau, E., Theraulaz, G., & Dorigo, M. (1999). *Ant Colony Optimization*. Oxford University Press.

Chan, F.T.S., & Tiwari, M.K. (2007). *Swarm Intelligence. Focus on Ant and Particle Swarm Optimization*. I-Tech Education and Publishing.

Dorigo, M. (1992). *Optimization, learning and natural algorithms*. Politecnico di Milano, Italy: Ph.D. Thesis.

Dorigo, M., & Stützle, T. (2004). *Ant Colony Optimization*. Cambridge: MIT Press.

Eberhart, R. C., Shi, Y., & Kennedy, J. (2001). *Swarm Intelligence: From*

Natural to Artificial Systems (Santa Fe Institute Studies in the Sciences of Complexity). Academic Press.

Engelbrecht, A. P. (2007). *Computational Intelligence: An Introduction.* John Wiley & Sons.

Kennedy, J., Eberhart, R. C., & Shi, Y. (2001). *Swarm Intelligence.* San Francisco. Morgan Kaufmann Publishers.

Krause, J., Cordeiro, J., Parpinelli, R. S., & Lopes, H. S. (2013). *Swarm Intelligence, and Bio-Inspired Computation. A Survey of Swarm Algorithms Applied to Discrete Optimization Problems.* Elsevier.

CHAPTER EIGHT
AI PLAYING GAMES

8.1 Introduction

This chapter introduces two major types of deterministic, no-chance AI games: single-agent games and multi-agent games. In single-agent games, the AI agent has to solve a jigsaw puzzle or find a way out of a maze almost instantaneously. There is no opponent to beat. In multi-agent games, the AI agent has to play against another agent or team of agents and try to win. Examples are tic-tac-toe, Othello, checkers, chess, or *Go*.

The Minimax algorithm is a fundamental multi-agent game playing AI algorithm. It analyzes a game between two players: *Max* who is trying to maximize his score and *Min*, who is trying to minimize the score of *Max*. The algorithm constructs a game tree and, aided by a heuristic function, determines *Max*'s winning moves. Recently, Machine Learning algorithms are also used in AI games. Section 8.3 explains reinforced learning used in training an agent to play the game of Othello.

In section 8.4, we will highlight the three most outstanding achievements of AI game playing programs in recent years: *Deep Blue*, *Watson*, and *AlphaGo* able to defeat world champions in chess,

Jeopardy, and *Go*, respectively. These AI successes have great future promise.

8.2 Game playing agents

This section analyzes the single-agent and multi-agent games in order to develop AI game playing algorithms and programs.

8.2.1 Single-agent games

In a single-agent game, the AI program is the sole agent in action with no opponent. The game playing AI agent is expected first, to solve the problem and second, to solve it optimally. The 8-puzzle is an example of a single-agent game (Fig. 8.1). We have seen in chapter 2 how the Best First Search algorithm solves the puzzle optimally. There is a major difference between humans and computers solving this problem. Humans will move the tiles by trial and error, looking at best only a few moves ahead. Humans cannot solve the puzzle by knowing the entire sequence of moves from the start to the goal. Computers, on the other hand, first execute the algorithm, obtain the entire sequence of moves as the solution and then manipulate the tiles.

Start state   Goal state

Fig. 8.1 **Single-agent game (8-puzzle)**

Another example is inputting the pieces of a jigsaw puzzle to the computer so that it can solve the problem by dovetailing the pieces together (Fig. 8.2). Here, we assume that the computer program has some visual aids for seeing the pieces of the jigsaw puzzle.

Fig. 8.2 **Single-agent game (jigsaw puzzle)**

8.2.2 Multi-agent games

In multi-agent games, two teams of players play against one another. The outcome of the game is a win, a loss or a draw as seen from the point of view of one team (Fig. 8.3). When there is a win for one team, it implies a loss for the opponent team and vice versa. In case of a draw, on the other hand, both parties obtain the same results.

Multi-agent game

8.3 Game playing algorithms

This section explains the Minimax game playing algorithm along with some of is variations.

8.3.1 The Minimax Algorithm

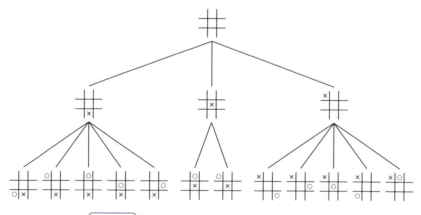

Fig. 8.4 **Partial game tree for tic-tac-toe**

Game playing algorithms rely on the game tree, which enumerates all the possible outcomes of a game. Fig. 8.4 shows a partial tree for the tic-tac-toe (noughts and crosses) game. Let us assume that the first player plays the crosses. Three unique configurations are possible at the starting level as shown in Fig. 8.4. The second level in the game tree shows the possible positions of the noughts. The tree unfolds the alternate playing positions of the two players till the end of the game. The biggest disadvantage of using a game tree is that it rapidly grows in depth and width.

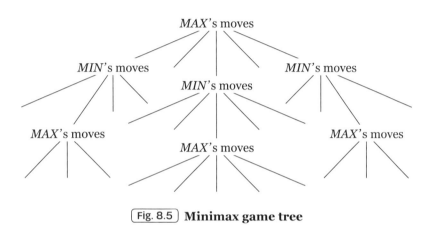

Fig. 8.5 **Minimax game tree**

The Minimax algorithm assumes there are two ideal players. One of the players is called *Max* and the opponent is called *Min*. *Max*, as the name implies, tries to maximize his score, while *Min* tries to minimize *Max*'s score. It is also assumed that both the players know all the outcomes of the game and that both play ideal games.

The algorithm creates a game tree, enumerating all possible moves of the two players at every turn. The Minimax game tree is shown in Fig. 8.5. In drawing the game tree, let us assume that *Max* has won the toss to begin play. The root node of the game tree indicates *Max*'s turn. The branches proceeding from the root node indicate all the possible moves *Max* can make. The next generation of nodes represents *Min*'s turn to play and the branches proceeding from each of these nodes indicate all the possible moves *Min* can make from that particular node. The game tree expands alternating between *Max*'s and *Min*'s turns till the end of the game. The steps of the Minimax algorithm are shown in Fig. 8.6 and an example is shown in Fig. 8.7.

Step 1: Draw the game search tree
Step 2: Determine *Max*'s and *Min*'s playing turns
Step 3: Give scores to each leaf node
Step 4: *Max*'s turn to play:
      Score = Max(Score of children nodes)
Step 5: *Min*'s turn to play:
      Score = Min(Score of children nodes)
Step 6: *Max*'s turn to play:
      Score = Max(Score of children nodes)
::::::::::::::::::::::::::::::::::::::
Step n: Draw arrows along the paths connecting chosen nodes

(Fig. 8.6) **The Minimax algorithm**

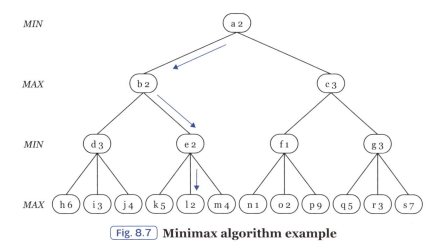

Fig. 8.7 Minimax algorithm example

Playing the Nim game using Minimax

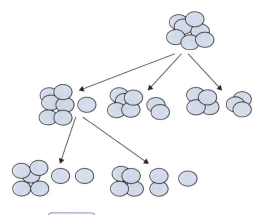

Fig. 8.8 The Nim game tree

The Nim game is a simple children's game rumored to have originated in Nepal. A pile of pebbles is placed between the two players. Each player takes turns in dividing any one of the pile of pebbles in *unequal* parts. The game ends when one of the players cannot further divide any of the resulting piles. This signifies a loss for the said player.

Fig. 8.9 shows the Minimax algorithm computing the winning strategy of Max.

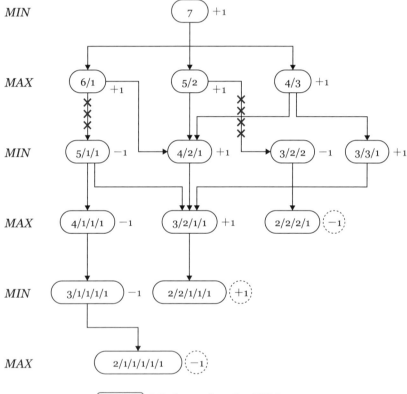

Fig. 8.9 Minimax for the NIM game

8.3.2 Alpha-beta pruning

Table 8.1 shows the game tree complexity of various board games. Traversing the entire game tree to find out the outcome of a chosen move (node) in real-time is not feasible even for a supercomputer. The α-β pruning technique helps reduce the search. This algorithm maintains two values: α, which represents the maximum score that the maximizing player is assured of and β, the minimum score that the minimizing player is assured of. Initially, α is negative infinity and β is positive infinity so that both players start with their lowest possible score. It can happen that when choosing a certain branch from a certain node the minimum score that the minimizing player is assured of becomes less than the maximum score that the maximizing player is assured of ($\beta \leq \alpha$).

Table 8.1   **Game tree complexity of various board games**

Game	State-space complexity	Game-tree complexity	Branching factor	Average game length
Tic-Tac-Toe	10^3	10^5	4	9
Connect Four	10^{13}	10^{21}	4	36
Othello	10^{28}	10^{58}	10	58
Checkers	10^{21}	10^{31}	-	-
Chess	10^{46}	10^{123}	35	80
Shogi	10^{71}	10^{226}	92	115
Go	10^{72}	10^{360}	250	150

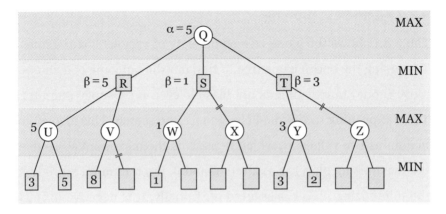

(Fig. 8.10) **Alpha-Beta pruning**

If this is the case, the parent node should not choose this node, because it will make the score for the parent node worse. Therefore, the other branches of that node need not be explored.

The α and the β values during the search are computed as follows:

The α value of a *Max* node is set equal to the current largest final backed-up value of its successors.

The β value of a *Min* node is set equal to the current smallest final backed-up value of its successors.

The search is discontinued under the following rules:

α *cut-off rule*: Search can be discontinued below any *Min* node having β value less than or equal to the α value of any of its *Max* node ancestors. The final backed-up value of this *Min* node can then be set to its β value.

β *cut-off rule:* Search can be discontinued below any *Max* node having an α value greater than or equal to the β value of any of its *Min* node ancestors. The final backed-up value of this *Max* node can then be set to its α value.

For example, starting at node Q (Fig. 8.10), the algorithm descends in depth-first order and computes the value of the evaluation function for each of the leaf nodes. These values are backed up to the parent node U (*Max* = 5). This is then assigned to the grandparent node Q as its β value. It cannot have a value larger than 5. The algorithm then descends to node Q's grandchildren and the search of their grandparent is terminated if any grandchild is greater than or equal to Q's β. This node V is β pruned. The algorithm continues pruning the tree with α and β cuts till the end of the right branch.

8.3.3 Artificial Neural Networks in game playing

Artificial Neural Networks (ANN) learning which consists in adjusting the connection weights is conventionally carried out by the Backpropagation Algorithm (BP). Recently, Evolutionary Algorithms and Swarm Intelligence Algorithms have also been used for ANN learning.

ANN have also been able to learn and excel in playing games without relying on any expert knowledge about how to play the game. Initially, the connection weights are randomly generated and some relevant information such as the board configuration, the number of pieces on the board, etc. is fed to the input layer of the ANN. The neurons in the ANN layers fire and give an output, which usually indicates the move to be made. Several agents (programs), each driven by an ANN, play against each other till they reach the end of the game. The winning agent is given a reinforcing reward, and the learning cycle is repeated. The learning agent improves its playing strategy with each learning cycle.

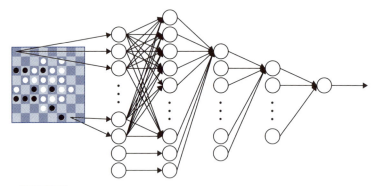

Fig. 8.11 **Neural network learning Othello wining moves**

For example, in the case of an agent learning to excel at the game of Othello, the board configuration, the piece differential and the legal moves from the current positon are fed to the input neurons in each learning cycle (Fig. 8.11). The ANN outputs the move the agent is to play next. The opponent is normally another agent driven by expert heuristics or by another ANN. After the response of the opponent, the agent calculates the next move as the output of the ANN. This sequence of actions continues till the end of the game. At the end of the game, the agent is given a positive score for win, a negative for loss and a zero for draw. This way of training the agent (ANN) by giving it a score depending on the game result is called reinforced learning. After each game, the connection weights of the ANN are re-adjusted by an Evolutionary or Swarm Intelligence algorithm to maximize the score. A large number of games are played in this way to attain the desired level of expertise.

8.4 Mega-successes of AI at playing games

The history of AI has gone through prolonged winters. However, towards the end of the 20th century, AI made great strides in several machine learning algorithms. The outstanding results could be seen in game playing as discussed in this section.

8.4.1 *Deep Blue* winning chess

In 1997, the IBM supercomputer *Deep Blue* defeated the chess world champion, Gary Kasparov (Fig. 8.12). *Deep Blue* decided its moves by computing the evaluation function, which is a measure of the goodness of a given chess position. The evaluation function considered four basic chess values: material, position, King safety, and tempo. Material is based on the worth of a particular chess piece. For example, if a pawn is worth 1 point, a knight or bishop 3, a rook 5 and the queen 9. However, it is impossible to conduct an exhaustive search to find the best possible move, because the game of chess has an enormous state-space of 10^{46} and game-tree complexity of 10^{123} (Table 8.1).

Instead of analyzing moves all the way to the end of the game, the chess-playing program can look only few moves ahead, but more than a human could manage. *Deep Blue* typically looked 12 plies ahead in all variations, generating around 200 million nodes per second. It would evaluate each of these node positions and add points for a range of positional factors, chosen from a data-base created by human grandmasters.

(Fig. 8.12) IBM's *Deep Blue* wins chess world championship

Deep Blue conducted the search using a massively parallel, RS/6000 SP Thin P2SC-based complete system with 30 nodes, with each node containing a 120 MHz P2SC microprocessor, enhanced with 480 special purpose VLSI chess chips. Its chess playing program was written in C and ran under the AIX operating system. In June 1997, *Deep Blue* was the 259th most powerful supercomputer according to the TOP500 list, reaching 11.38 GFLOPS on the High-Performance LINPACK benchmark.

8.4.2 *Watson* winning Jeopardy

In 2011, IBM's supercomputer *Watson*, named after IBM's founder, defeated the reigning Jeopardy champions, Brad Rutter and Ken Jennings (Fig. 8.13). Jeopardy is a well-known TV quiz show in the US. Rather than answer questions as done in normal quiz shows, the Jeopardy contestants are provided with a few clues, which they have to glue together to frame a question. This meant that *Watson* had to take the clues from the quiz, do natural language processing and frame the Jeopardy questions.

Fig. 8.13 **IBM's *Watson* wins the game of Jeopardy**

Clues were provided to *Watson* as text messages via a typewriter as in the classic Turing Test (explained in chapter 10).

The software components and the data contents used inside *Watson* are shown in Table 8.2. *Watson* uses a complex combination of natural language processing, semantic analysis, information retrieval, automated reasoning and machine learning to answer the questions. *Watson* uses many existing algorithms to generate potential answers. First, the sentence is parsed. Then hypotheses are created. These hypotheses are then checked against evidence. Finally, the hypotheses have confidence levels assigned to them. If the top hypothesis has a confidence level above the threshold, *Watson* proposes an answer. All this is done under 3 seconds in the actual game of Jeopardy.

Table 8.2 *Watson*'s software components and data contents

Software components	Structured text	Unstructured text
SUSE Linux (OS)	WordNet	Encyclopedias
Apache Hadoop	DBPedia	Wikipedias
Apache UIMA	YAGO	Dictionaries & Thesauri

The parallel processing inside *Watson* is performed by specially designed hardware, consisting of 90 IBM Power 750 Servers with a total of 2,880 POWER7 processor cores and 16 terabytes of RAM. The software components include Apache Hadoop and UIMA (Unstructured Information Management Architecture). UIMA provides standards-based frameworks that allow analysis and annotation of large volumes of computer text. *Watson* used Apache UIMA for real-time content analytics and natural language processing.

8.4.3 *AlphaGo* winning *Go*

Go is a 2,500-year-old game very popular in China, Korea, and Japan. *Go* is played with black and white game stones on a 19 x 19 grid board. Players take turns to place a stone on the board, attempting to surround some of the opponent's pieces. The goal of the game is to surround the largest area of the board with one's pieces.

Fig. 8.14 **Google's *AlphaGo* wins the *Go* championships**

Go is a sophisticated game requiring creativity and intuition, so much so that experts had predicted it would take decades of technological advance for AI programs to win against human *Go* champions. However, all the expert opinions were shattered in March 2016, when the Google DeepMind supercomputer *AlphaGo* became the first game playing AI program to win a *Go* match against Lee Sedol, the 9 dan *Go* world champion. In the widely telecast match played in Seoul from March 9 to 15, 2016, *AlphaGo* won 4 out of 5 games (Fig. 8.14).

The *AlphaGo* algorithm is a combination of the following two game playing strategies:

Monte Carlo Tree Search

Any simulation performed with random numbers is called Monte Carlo simulation (named after the famous gambling venue in Monaco). Recall that the game tree is a colossal tree with massively large breadth and depth. Exhaustive search in reasonable time is impossible even for supercomputers. While the α-β pruning systematically prunes branches which are judged to be unprofitable, Monte Carlo Tree Search chooses branches at random and then simulates the game to the very end to find a winning strategy.

Deep Learning

As we have seen in chapter 5, the Deep Learning method employs convolutional nets containing a large number of layers. Among other layers with millions of neuron connections, the *AlphaGo* AI algorithm contains a policy network that selects the next move and a value network that predicts the winner of the game. This is then backtracked to the most appropriate move.

Artificial Neural Networks (ANN) are generally trained on a training dataset and then tested on an entirely different test dataset. If the ANN shows good prediction results when training on the training set, but performs poorly on the test dataset, then we say that the ANN are over-trained. They have picked up too many peculiarities of the training dataset (including noise) and have not learnt enough to extract the general characteristics of the data. Successful performance on the test dataset is an indication of generalized learning and pattern recognition. The learning success of ANN was amply demonstrated by *AlphaGo*. It had trained on millions of games played by *Go* professionals and further perfected its strategy by playing a large number of games against itself. It was not, however, exposed to a single game played by Sedol Lee in the training phase. Nevertheless, it won against the world champion by a big margin in the very first face-to-face challenge. At the end of the match, many experts commented that *AlphaGo* demonstrated flexibility and creativity in its moves, features not seen in hard-wired game playing programs.

8.5 Exercises

1. Define the heuristic function for the tic-tac-toe game.
2. Draw a partial game tree for playing tic-tac-toe. Use the Minimax algorithm to decide the winning moves.
3. Define the heuristic function for the Othello game.
4. Define the heuristic function for the mini (4 x 4) Othello game. Draw a partial Minimax tree.
5. Can you suggest an algorithm to put together a jigsaw puzzle?
6. Think of a simple board game and work out Minimax.
7. Could *Watson* use its natural language processing capabilities to hold a normal conversation with a human?
8. Could *Watson* do spontaneous machine translation?

9. Could *Watson* play games and win against human champions?
10. Could you train *Deep Blue* to play *Go* and *AlphaGo* to play chess?

References

Baker, S. (2012). *Final Jeopardy: The story of Watson, the computer that will transform our world*. Boston, MA: Mariner Books.

Fogel, D. B. (2002) *Blondie24: Playing at the Edge of AI*. San Francisco, CA: Morgan Kaufmann.

González Calero, P. A. (2011). *Artificial Intelligence for computer games*. New York: Springer.

Hsu, F-H. (2004). *Behind Deep Blue: Building the computer that Defeated the world chess champion*. Princeton, NJ: Princeton University Press.

Kasparov, G. (2017). *Deep Thinking: Where Machine Intelligence Ends and Human Creativity Begins*. Public Affairs.

Millington, I., & Funge, J. (2009). *Artificial Intelligence for Games*. (2nd ed.). CRC Press.

Muthusamy, D. (2016). *Neural Network Architectures for Solving Imperfect Information Game*. Lap Lambert Academic Publishing.

Newborn, M. (2003). *Deep Blue: An Artificial Intelligence Milestone*. New York, NY: Springer.

Zhou, Y. (2017). *AlphaGo vs Lee Sedol: The Match that Changed the World of Go*. CreateSpace Independent Publishing Platform.

## CHAPTER NINE

LIFE IS A GAME

9.1 Introduction

Most of us would think it is a pun to say, "Life is a Game." What if we were to discover that life which we respect and marvel at is only a game played by tiny microscopic biological units called cells! In this chapter, we shall be considering two *artificial* phenomena: Cellular Automata and Game of Life. Starting from very simple configurations and following a set of very simple rules, these phenomena evolve into unpredictably complex configurations, which appear to be "living, growing and dying."

Cellular Automata (CA) are a group of cells placed on an infinite grid. The cells are each initially in a state of 1 or 0. Depending on the state of the neighboring cells, the automata evolve through several discrete time-steps that follow a set of predefined rules. The myriads of patterns into which they evolve have practical applications.

The Game of Life (GOL), often referred to as Life, invented by J. H. Conway in 1970, is an extension of the CA. GOL is not really a game, for there are neither players nor winning and losing in this game. Rather, GOL consists of a set of simple cells that evolves from an initial configuration into patterns of cells on an infinite grid, guided by a handful of primitive rules. This chapter looks into GOL config-urations such as still forms, oscillators, gliders, spaceships, and pentaminoes.

9.2 Cellular Automata

An automaton (plural: automata) refers to a machine, or any mech-anism that automatically follows a predetermined sequence of self-guided operations. CA are automata consisting of cells. This section describes CA consisting of one-dimensional and two-dimen-sional arrangements of cells.

9.2.1 Elementary Cellular Automata

An elementary CA is a one-dimensional array of cells, each of which can be in either state 1 or 0. The change of state of a cell in the next generation is determined by the current state of that particular cell and its immediate neighbors. The cell with its immediate neighbor to the right and to the left form a group of three. Since each one can be either in state 1 or 0, a total of 2^3 (=8) states are possible. Fur-ther, there are a total of 2^8 (=256) rules defining the resulting state for each possible configuration. Fig. 9.1 shows CA rule 30 (30 is the decimal equivalent of the binary number 00011110).

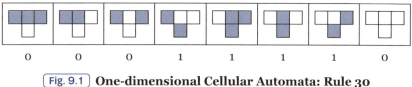

0   0   0   1   1   1   1   0

Fig. 9.1 One-dimensional Cellular Automata: Rule 30

The evolution of a one-dimensional cellular automaton proceeds as follows: the zero[th] generation cell is placed on the first row of the grid. The state of the automaton is then evolved according to the prescribed rule and the resulting second generation configuration is placed on the second row, and so on. For example, Fig. 9.2 illustrates a pattern generated by the first 250 generations of rule 30 one-dimensional CA starting with a single black cell.

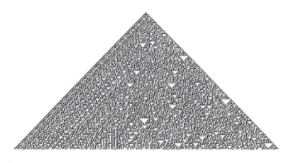

Fig. 9.2 **CA pattern of 250 generations according to Rule 30**

Some rules evolve into arbitrary patterns, while others into amazing symmetrical patterns such as those shown in Fig. 9.3.

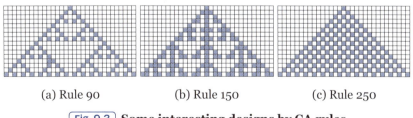

(a) Rule 90   (b) Rule 150   (c) Rule 250

Fig. 9.3 **Some interesting designs by CA rules**

Rule 30 patterns

Rule 30 is a one-dimensional cellular automaton. Owing to its intriguing nature, Rule 30 has been subjected to extensive research. Being aperiodic and chaotic, Rule 30 produces amazingly complex patterns some of which are found in nature. For example, the colorful design on the shell of the cone snail species, *Conus textile*, bears a striking resemblance to the patterns generated by Rule 30 (Fig. 9.4).

Fig. 9.4 **Shell patterns of *Conus textile*, similar to Rule 30**

9.2.2 Two dimensional Cellular Automata

Two dimensional CA are a spatial lattice of n cells, each of which is in one of k states at time t. Each cell follows the same simple rule for updating its state. The cell's state at time $t+1$ depends on its own state and the states of some number of neighboring cells at t. CA neighborhoods are of three types (Fig. 9.5):

(a) Von Neumann: Only the four cells each sharing one of the four edges of the cell count as its neighbors.

(b) Moore: The four cells sharing the edges of the cell as well as the remaining four other cells touching the four corners form the neighborhood.

(c)　Extended Moore: When we include the external layer of cells, which are in contact with the Moore neighborhood, we get the Extended Moore neighborhood.

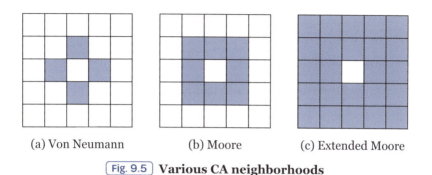

(a) Von Neumann　　　(b) Moore　　　(c) Extended Moore

Fig. 9.5　**Various CA neighborhoods**

9.2.3 Cellular Automata applications

Cellular automata produce complex structures from simple rules. Although an abstract mathematical concept, CA can be applied to perform practical simulations of the spread of epidemics, social phenomena, predator-prey models, and the spread of forest fires. CA rules can also be used to explore complex natural and biological phenomena. Since most natural phenomena contain randomness, we need random number generator algorithms to simulate them on computers. Rule 30 is a good candidate for random number generation. Random simulations performed on a computer are called *Monte Carlo* simulations. These simulations generate (pseudo) random numbers, test their degree of randomness statistically, and use them extensively to perform large scale simulations. Another prominent area of random number applications is *Cryptography* (encoding and decoding of information for computer security).

9.3 Game of Life

Game of Life (GOL) is not a game in the conventional sense, because there are neither players nor winning and losing. Rather, it refers to a kind of simulation in which cells placed in some arbitrary environment evolve in discrete time-steps leading to some interesting and useful patterns. The environment in which each cell finds itself is crucial in deciding the fate of that cell.

9.3.1 GOL rules

This section describes the environment, the neighborhood and the rules of GOL.

The environment

The environment of the cell is an infinite plane or grid containing other cells. In the initial configuration, each cell is either alive or dead. It is customary to depict the live cells in color and the dead cells in white. Different authors use different color conventions. We shall follow the color conventions shown in Fig. 9.6.

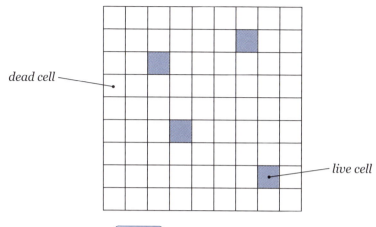

Fig. 9.6 GOL environment

The neighborhood

The cells in the immediate proximity of a given cell become its neighbors. For example, in Fig. 9.7, neighbors of 1 are 8, 9, and 2; neighbors of 17 are 9, 10, 11, 16, 18, 23, 24, and 25. GOL neighborhood is the same as the Moore neighborhood discussed in section 9.2.

1	2	3	4	5	6	7
8	9	10	11	12	13	14
15	16	17	18	19	20	21
22	23	24	25	26	27	28
29	30	31	32	33	34	35
36	37	38	39	40	41	42
43	44	45	46	47	48	49

Fig. 9.7 GOL neighborhood

A single plane is bounded on four sides. Therefore, the neighbor-hood of the cells outside the boundaries are not defined. The single plane can be rolled along its length to form a cylinder and then the two ends of the cylinder rolled and sealed together to form a toroid. The cells on the toroid are seamlessly connected without any bound-ary, enabling us to define the neighborhood for every cell on the to-roidal surface (Fig. 9.8).

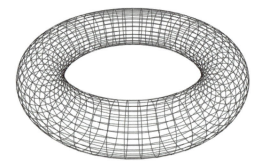

Fig. 9.8 **Toroid: Cells connected seamlessly without boundary**

Rules of GOL

The GOL evolves with the following simple rules:

(1) A living cell *continues to live* in the next time-step if it has two or three live neighbors.

(2) A dead cell *becomes alive* in the next time-step if it has ex-actly three live neighbors.

(3) A living cell *dies* in all other cases.

According to rule (3) when the live neighbors are less than two, the cell dies due to *loneliness*. When the live neighbors are more than three, the cell dies due to *over-congestion*.

The contribution of the neighbors to the cell in its evolutionary history is show in Table 9.1.

Table 9.1 **Rules of the Game of Life**

Number of neighbors	Cell state	Contribution by the neighbors
0, 1	dies	loneliness
2	stays alive	support
3	new birth	feed
4 +	dies	over-congestion

Refer to the example of the GOL grid shown in Fig. 9.9. Cells 13, 17, 18, 24, 25, 32, and 49 are alive at the current time-step. In the next time-step, cells 13 and 49 will die of isolation, since they are not surrounded by live neighbors. Cells 24 and 25 will die of over-congestion since each of them is surrounded by four live cells. Cell 31 will become alive since it has exactly three live neighbors, and cells 17 and 18 will continue to live, each being surrounded by exactly three cells. Cell 32 will also survive because of two live neighbors.

1	2	3	4	5	6	7
8	9	10	11	12	**13**	14
15	16	**17**	**18**	19	20	21
22	23	**24**	**25**	26	27	28
29	30	31	**32**	33	34	35
36	37	38	39	40	41	42
43	44	45	46	47	48	**49**

1	2	3	4	5	6	7
8	9	10	11	12	13	14
15	16	**17**	**18**	19	20	21
22	23	24	25	26	27	28
29	30	**31**	**32**	33	34	35
36	37	38	39	40	41	42
43	44	45	46	47	48	49

(a) At time = t   (b) At time = t+1

Fig. 9.9 GOL example

Properties of GOL

GOL exhibits the following properties:

- Localism: States are updated based on the properties of the neighborhood.
- Parallelism: The state of each and every cell is updated at the same time in parallel.
- Homogeneity: The same set of rules is applied to each of the cells.

9.3.2 GOL patterns

GOL develops phenomenal patterns when allowed to evolve on its own, following the predefined simple rules. Some of these patterns are mentioned in this sub-section.

Still Life forms

Still Life refers to a pattern that does not change from a given generation to the next. A variety of still Life patterns are produced from initial random patterns. Some of them are described below.

Block

The block consisting of four live cells is a basic still Life pattern. Since each living cell has exactly three live neighbors, it continues to live in the next generation. The block is surrounded by twelve dead cells.

Fig. 9.10 **Block**

Since none of these dead cells has more than two live neighbors, they continue to be in the dead state from one generation to the next. The live block of cells bounded by a rim of dead cells persists indefinitely as still Life (Fig. 9.10).

Boat

Fig. 9.11 shows the structure of a boat formed by five cells. The live cells in the bow as well as in the stern of the boat have exactly two live neighbors to enable them to carry on living in the succeeding generations, while the dead cell in the middle continues to remain dead being stifled by the surrounding live cells.

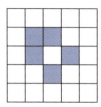

Fig. 9.11 **Boat**

Ship

The symmetrical structure of six live cells formed by three in front and three at the back with a central dead cell resembles a ship. The live cells continue living in the succeeding generations due to the presence of two or three live neighbors, while the central cell continues in the dead state being overcrowded by live neighbors (Fig. 9.12).

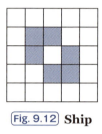

(Fig. 9.12) **Ship**

Beehive

The beehive (Fig. 9.13) pattern consists of live cells enclosing two dead cells. Each of the six live cells has exactly two live neighbors, so that it continues to survive in the following generation. On the other hand, each of the two dead cells in the middle is surrounded by five live cells. Therefore, both remain dead due to overcrowding.

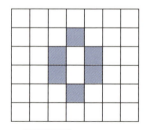

(Fig. 9.13) **Beehive**

Loaf

The loaf (Fig. 9.14) has three dead cells at the center surrounded by a rim of live cells. None of the cells can change its states.

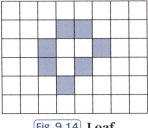

Fig. 9.14 **Loaf**

Periodic Life forms

Some Life forms repeat themselves after a fixed number of generations called its period. These periodic Life forms are known as oscillators because they appear to be oscillating to and fro on the Life grid. Some of the oscillators are:

Blinker

The blinker (Fig. 9.15) consists of three live cells in a row. It alternates from a horizontal to a vertical position and gives the impression of blinking.

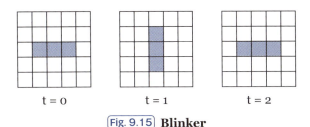

t = 0   t = 1   t = 2

Fig. 9.15 **Blinker**

Toad

This configuration (Fig. 9.16) hops like a toad on the grid.

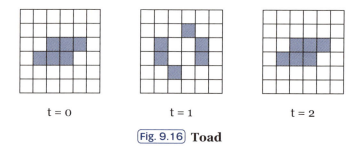

t = 0   t = 1   t = 2

Fig. 9.16 **Toad**

Beacon

The beacon (Fig. 9.17) gives the impression of a beam of light flashing. The blinker, toad, and beacon all have an oscillation period of 2 time-steps.

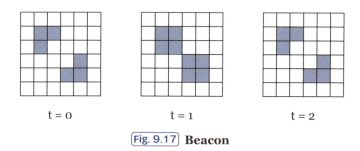

t = 0   t = 1   t = 2

Fig. 9.17 **Beacon**

Pulsar

The pulsar (Fig. 9.18) configuration pulsates on the grid with an oscillation period of 3.

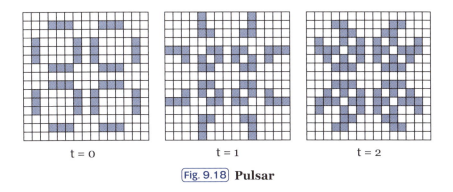

| t = 0 | t = 1 | t = 2 |

Fig. 9.18 **Pulsar**

Moving Life forms

The moving Life forms are patterns on the grid that keep moving from generation to generation. A finite pattern may return to its initial state after a number of generations, but in a different location. The number of generations in which the pattern repeats itself is called its period. Some moving Life forms appear like spaceships. The speed of Life spaceships is given by:

$$v = \frac{nc}{p} \qquad (9.1)$$

where, c refers to the movement of one cell in one generation,
(c is the speed of light in Einstein's theory of Relativity),
n is the number of cells the pattern has moved across the grid, and
p is the period or the minimum number of generations it takes the spaceship to appear in a different location.

Glider

The glider is a simple 5-cell pattern easily produced from randomly generated initial configurations. It is the smallest and the most common lightweight spaceship travelling diagonally across the Life grid (Fig. 9.19). Several gliders are often used in GOL so that by colliding with one another more complex patterns are produced.

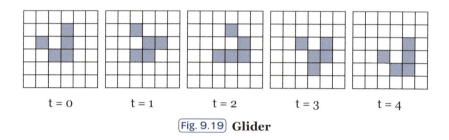

t = 0   t = 1   t = 2   t = 3   t = 4

Fig. 9.19 **Glider**

Conway discovered three kinds of spaceships: Light Weight Space Ship (LWSS), Medium Weight Space Ship (MWSS), and Heavy Weight Space Ship (HWSS) (Fig. 9.20).

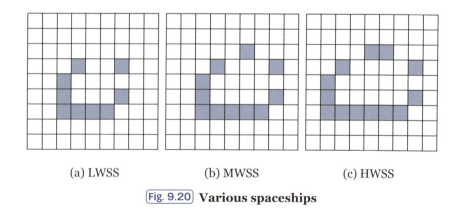

(a) LWSS   (b) MWSS   (c) HWSS

Fig. 9.20 **Various spaceships**

Table 9.2 **Speeds of various moving Life forms**

Life form	n	p	v
Glider	1	4	c/4
LWSP	2	4	c/2
MWSP	2	4	c/2
HWSP	2	4	c/2

Tagalongs are small patterns that are attached to the back of a spaceship and move with it. A pushalong is a tagalong attached to the front of the spaceship. Debris are patterns left behind by spaceships as they move along. Table 9.2 lists the spaces of various moving Life forms and Table 9.3 lists the evolving patterns produced by an initial pattern consisting of k number of live cells.

Table 9.3 **Evolving patterns of various initial configurations**

Initial pattern with k live cells	Evolving pattern
1	disappears
2	disappears
3	becomes blinker
4	becomes beehive at t=2
5	becomes traffic lights at t =6
6	disappears at t =12
7	becomes honey farm
8	becomes blocks and beehives
9	becomes two sets of traffic lights
10	becomes pentadecathlon
11	becomes 2 blinkers
12	becomes 2 beehives

9.3.3 Pentominoes

In mathematics, a pentomino is a pattern formed by arranging five (Greek, *pente* = 5) blocks side by side. It may be looked upon as an arrangement of five dominoes. In all, twelve distinct configurations are possible, if we leave out the rotational and the translational symmetries. Conway named the configurations O, P, Q, R, S, T, U, V, W, X, Y, Z since they somewhat resemble these letters. The new naming scheme for all the twelve configurations is shown in Fig. 9.21. Re-arrange the configurations to form two words: "FLIPS" and the last series of letters arranged in order from T to Z.

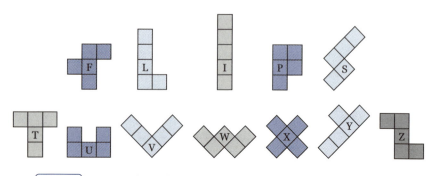

(Fig. 9.21) **Pentamino distinct configurations: "FLIPS" & "T-Z"**

R-pentomino

The evolving configurations of pentaminoes have been extensively studied in GOL scenarios. Most of the pentaminoes stabilize in less than 10 generations. However, the evolution of the R-pentamino (F-pentamino in the new naming scheme) is remarkable. It contin- ues growing for 1103 generations and at intermediate generations produces well-known patterns like blocks, blinkers, traffic lights, boats, beehives, and toads. The stable pattern produced at generation 1103 has 116 cells. At this stage the identifiable pattern consists of eight blocks, six gliders, four beehives, four blinkers, one boat, one loaf, and one ship.

9.4 Exercises

1. Use CA to simulate a predator-prey model.
2. Use CA to simulate a forest fire.
3. Play the Game of Life on a hexagonal grid.
4. How would you design an algorithm to determine whether the initial pattern of GOL has not stopped evolving?
5. Change the basic rules of GOL and work out new patterns.
6. Change the neighborhood definition and work out GOL patterns.
7. Introduce randomness in the rules of GOL and work out the evolution.
8. Work out all the symmetrical arrangements of pentominoes.
9. Create animal shapes by fitting several pentaminoes together as shown in Fig. 9.22.

Fig. 9.22 **Pentamino puppy**

10. According to Goldbach's conjecture, every even number can be expressed as the sum of at least one pair of prime numbers. Use CA to find the constituent pair(s) of prime numbers.

References

Adamatzky, A. (Ed.) (2010). *Game of Life Cellular Automata*. (1st ed.). London: Springer.

Adamatzky, A., & Martínez, G. J. (2016). *Designing Beauty: The Art of Cellular Automata*. Springer.

Batty, M. (2007). *Cities and Complexity: Understanding Cities with Cellular Automata, Agent-Based Models, and Fractals*. MIT Press.

Codd, E. F. (2014). *Cellular Automata*. Academic Press.

Hoekstra, A. G., Kroc, J., & Sloot, P. M. A. (Eds.). (2010). *Simulating Complex Systems by Cellular Automata*. Springer.

Ilachinski, A. (2001). *Cellular Automata: A Discrete Universe*. World Scientific Publishing Company.

Li, T. M. (2011). *Cellular automata*. New York: Nova Science.

McIntosh. H. V. (2009). *One Dimensional Cellular Automata*. Luniver Press.

Schiff, J. L. (2008). *Cellular automata: A discrete view of the world*. Hoboken, NJ: Wiley-Interscience.

Weimar, J. (2003). *Simulation with Cellular Automata*. (2nd ed.). Logos Verlag.

Wolfram, S. (2002). *A New Kind of Science*. Wolfram Media.

CHAPTER TEN
ARTIFICIAL SUPERINTELLIGENCE

10.1 Introduction

We have seen quite in detail the success of AI in limited areas like Search, Expert Systems, Fuzzy logic, Machine Learning, Metaheuristic Optimization, and Game Playing. Do AI programs really *"think and solve the problems as we humans do?"* There are serious objections to the central claim of AI experts that machines are capable of thinking. These objections are considered in section 2.

Human intelligence is tested by the Intelligence Quotient (IQ) test. If a future machine ever claims to be intelligent, how shall we test its intelligence? Turing himself suggested the famous *Turing Test*, various versions of which are explained in section 3. Robots are an extended form of AI and different kinds of robots and the laws of robotics proposed by experts to regulate their behavior in human society are presented in section 4.

AI enthusiasts and critics alike are awestruck at the rapid rate of growth in hardware and software systems. They foresee the approaching *Singularity*: a point in the not so distant future, when AI will suddenly acquire critical mass, get out of human control and become Artificial *Superintelligence*. The arguments for and against such a possibility and its social repercussions are presented towards the end of this chapter.

10.2 Objections to machines thinking

"Machines cannot think, and they never will!" is a remark we hear very often. The following sub-section presents some of the major objections to the idea that future machines will be able to think.

10.2.1 Lady Lovelace objection

Ada Lovelace, the daughter of Lord Byron is regarded as the first computer programmer. The programming language *Ada* is named after her. From her programming experience, Ada Lovelace concluded that computers do only exactly what they are told to do by the instructions in the program. This argument holds even today, for even the most powerful computers solving complex problems do nothing more than faithfully execute the instructions in the code, albeit at an extremely high speed.

The story, however, is different in the case of the AI programs. These programs modify and optimize the very algorithm which they execute. It is customary to refer to these AI programs as "agents" because they are autonomous and make their own decisions. In game playing, only a basic set of the game rules are coded in the agent program. The agent learns the finer rules of the game, the strategy of the opponent, and plans its own winning strategy. Game designers and programmers are often amazed at the self-learned creativity and unorthodox playing style of some AI programs.

10.2.2 The Chinese Room Argument

The Chinese Room Argument is a thought experiment proposed to demonstrate that the so-called intelligent computers do not understand at all what they are doing. Let us say, a person who has no knowledge of languages other than English is locked up in a room. All he has with him is a Chinese-English and an English-Chinese dictionary. People from outside the room ask him text-based questions in Chinese. He takes the Chinese characters (symbols) looks them up in his dictionaries, frames his responses in Chinese and sends them to the people outside. The people outside become convinced that the English speaker locked up in the room *understands* Chinese. But we the experimenters know that the English speaker gives correct answers to the Chinese questions *without really understanding* Chinese.

Now let us replace the English speaker in the thought experiment with a computer. The computer receives the questions in Chinese, does a look-up in its internal dictionaries (databases) and responds to the queries in Chinese, without really understanding a single Chinese character. Although the people outside the room are convinced that the computer understands Chinese and responds to them in Chinese, all that the computer does is: take the given Chinese symbols, consult a look-up table and produce other Chinese symbols as output.

The Chinese Room Argument demonstrates that computers manipulate symbols without understanding their meanings. Hence, computers are not intelligent like humans who *understand* the symbols they are manipulating.

10.2.3 Emotional Intelligence objection

Humans have not only a rational intelligence, traditionally measured by an IQ test, but they also have an emotional intelligence that psychologists attempt to measure by the EQ (Emotional Quotient) test. Human creativity like writing poetry or novels, painting, composing music or loving other humans, animals and nature arises from the Emotional Intelligence. Creativity, in particular, is a unique feature of human intelligence that cannot be replicated by machines. Some linguists assert that language is unique to human beings owing to the biological nature of the neurons. It seems that creativity and language are so innate to humans that they cannot be imitated by synthetic brains.

10.2.4 Philosophical objection

The ancient Greek philosopher Plato was probably the first to make a distinction between matter and spirit. In his famous work, *Timaeus*, the human person is made up of a material part and a non-material part. The material part is the body and the non-material part is the spirit, soul or mind. The basic distinction between matter and spirit is that the former is *composite* (made up of parts and hence can disintegrate), while the latter is *simple* (does not contain any parts and hence cannot disintegrate). Thoughts, memories, and emotions reside in the non-material spirit and not in the material brain. The spirit makes use of the material brain to exercise the functions of thinking, remembering, reasoning, and willing. The material brain by itself cannot perform any of these intellectual functions.

The French mathematician and philosopher, Descartes revived the Platonic material/spirit dualistic philosophy in the 17th century. This view of the spirit residing in the material human body and carrying on all the intellectual functions like thinking, reasoning, recollecting, feeling, and willing is often criticized as the "ghost in the machine" philosophy.

The Good Old Fashioned AI (GOFAI) tried to imitate the human mind as a whole. Today's AI has long abandoned this utopia. Instead, it has adopted the standard approach of designing systems that work in an intelligent way, evaluated from their external behavior and output rather than from their inner working. The basic assumption of today's AI is that the brain is an extremely complex electric computer and that intelligence is *computational*. Since the electrical activity of the neurons in the brain is reproducible in the laboratory, the fact that the functions of the brain are electrical in nature is established beyond doubt. Most AI researchers hold that it is only a question of time before computers will demonstrate capabilities of language and creativity.

10.3 Testing the intelligence of machines

This sub-section deals with the esoteric subject of testing machine intelligence.

10.3.1 Tesler's Theorem

Very often, AI researchers find themselves in puzzling situations. Critics present a difficult problem to challenge AI to prove its intelligence. AI researchers struggle with the problem for decades and one fine day come up with a solution. But the moment the problem is solved by AI, it is no longer an intelligent problem in the eyes of the critics and of the general public. This frequent shift in what is perceived as "intelligent" by the general public is referred to as Tesler's theorem.

In the days before the advent of electronic computers, people who were highly skilled in complex math calculations were called "computers." These computers were not ordinary, but highly intelligent people, and, their calculating jobs were considered intelligent jobs. Then came the electronic computers that managed to do much more complex calculations in a fraction of the time that the human computers required. But in public opinion, the electronic computers were only doing "calculations." They were not intelligent.

Playing chess is considered to be a mark of human intelligence. Human chess players who achieve grandmaster levels are often called prodigies, geniuses, super-intelligent. Called the "drosophila of AI," chess had been a challenge for AI from its inception. For decades, AI chess playing programs were ridiculed as rigid, brittle, and amateurish. Grandmasters considered it below their dignity to play against such amateur programs. In 1997, defying all the predictions to the contrary, IBM's chess playing supercomputer *Deep Blue* toppled the reigning world champion, Garry Kasparov. Although, Kasparov himself admitted that he felt he was playing against an alien kind of intelligence, most people immediately explained the match away as being played by a machine which does no more than number crunching. "Deep Blue winning at chess is no more impressive than a bulldozer winning an Olympic gold medal in weight lifting!" is another favorite quote of the AI critics.

10.3.2 Turing Test (TT)

Will future computers surpass human intelligence? This question was first proposed by Alan Turing, considered the father of machine intelligence. In his 1950 seminal paper, he proposed a test, known after him as the Turing Test (TT). A human judge using a computer monitor, keyboard, and mouse administers the test. The computer whose intelligence is to be judged is hidden together with a human person behind a screen (Fig. 10.1). The judge inputs a series of text-based questions through the keyboard. Some questions are answered by the computer and others by the human person. If at the end of the question-answer session, the judge cannot distinguish between the computer and the human answers, he declares the computer intelligent. In other words, if the computer behind the screen succeeds in deceiving the judge that it is human, it passes the Turing Test.

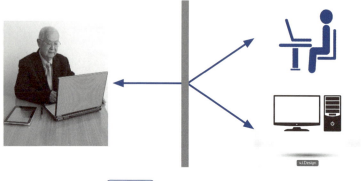

Fig. 10.1 **The Turing Test**

10.3.3 Total Turing Test (TTT)

The Turing Test applies only to the verbal behavior of computers, not using audible speech, but just a text typewriter. Many critics point out that this is not sufficient to test general intelligence. In the Total Turing Test (TTT), verbal as well as other behaviors are used to test the intelligence of machines. To pass the TTT, an intelligent candidate must be able to do, in the real world of objects and people, everything that real people can do. It must be capable of demonstrating intelligence at the sensory and motor levels. To pass the TTT, the computer must, among other actions, perceive objects (*computer vision*), move around them and manipulate them (*robotics*).

10.3.4 The Loebner Prize

In 1990, Hugh Loebner along with The Cambridge Center for Behavioral Studies proposed a contest designed to implement the Turing Test. The contest consists of 4 rounds involving 4 judges. In each round, each of the judges interacts with the two remotely hidden entities using a computer terminal. One of these entities is human confederate and the other an AI system. After 25 minutes of questioning, the judge must decide which entity is the human and which is the AI. If a system can fool half the judges that it is human under these conditions, a silver medal and a cash prize of $25,000 are awarded to the creator of that AI system.

The 2016 Loebner Prize contest[1] was held at Bletchley Park (where Alan Turing worked as a code-breaker during World War II) on 17 September 2016. After 2 hours of judging, the first prize was awarded to the creator of the AI system, *Mitsuku*, that scored 90/100.

10.4 Robots

Robots are devices that perform automated tasks. In chapter 1, we defined Robotics as the science of making extended AI, intelligent programs or agents clothed in a mechanical body. The extended body contains sensors to perceive the environment and actuators to act on the perceived environment. This section describes the different types of robots and the rules governing their behavior.

---

1   http://www.aisb.org.uk/events/loebner-prize (accessed in March, 2017).

10.4.1 Robot types

Robots could be pure mechanical devices, pure software agents or a combination of software driving hardware. These three categories of robots are described below.

Mechanical robots

The robotic arms that we see in production lines in an industrial setup are not intelligent. No learning ability is needed, because they are working in a highly constricted non-changing environment. Such mechanical robots are pre-programmed. They do exactly what they are told to do. Lady Lovelace would be right in judging these kinds of robots (and computers) as not intelligent.

Web robots

Web robots (WWW robots or Internet robots), or simply *bots*, are AI programs extended to the World Wide Web. They are designed to automate repetitive tasks on the Internet. One of the most common tasks of bots is web crawling, by which they fetch, analyze and file information from web servers. The Internet now is flooded with an increasing variety of bots: weather bots, game bots, advertising bots, auction bots, and chat bots, each performing a specified task autonomously and by interacting with other bots. Very soon, bots will become our personal assistants on the Internet. They will receive messages from us, perform a task on our behalf, and respond just as a human would.

Unfortunately, there are malicious bots that can play havoc with Web services. A botnet (combination of the words robot and network) is a group of Internet-connected computers engaged in interrupting Internet service. Spambots bombard ordinary users with a large number of spam mails. Fraud bots generate false information and false clicks to obtain financial gains for their creators. Chatterbots camouflage themselves as persons and obtain credit card numbers and other private information from unsuspecting victims. Keeping a check on the activities of malicious bots is becoming day by day more difficult.

Humanoid robots

Humanoid robots look like humans in their external appearance and behavior. Recent ones developed in Japan have amazing skills such as playing musical instruments, understanding and responding in human language, and guiding customers in restaurants and shopping malls (Fig. 10.2). Present-day humanoid robots are no more than Expert Systems with flexible arms and limbs. They are designed with the aim of performing dedicated tasks in practical, but narrow and confined domains. A violinist robot, for instance, might be a virtuoso in playing the *Devil's Thrill Sonata*, but not be able to deliver pizza at your door. Future humanoid robots should be the best candidates to face the challenge of the Total Turing Test to demonstrate their skills and intelligence over a range of activities performed in real-life environments.

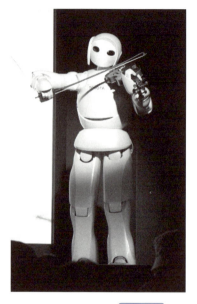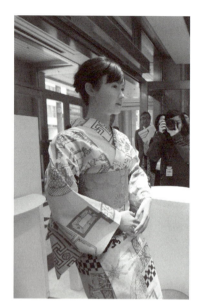

(Fig. 10.2) **Humanoid robots**

10.4.2 Asimov's Laws of Robotics

Robots are developed with the intention of performing tedious menial tasks so as to serve humanity. However, there are concerns that they may harm their human masters. To avoid such a catastrophe, the sci-fi writer Isaac Asimov has framed three laws of robotics, stated below:

1. A robot may not injure a human being or, through inaction, allow a human being to come to harm.

2. A robot must obey the orders given it by human beings except where such orders would conflict with the First Law.

3. A robot must protect its own existence as long as such protection does not conflict with the First or Second Laws.

10.5 Unprecedented growth of AI

Technology has been advancing at a rapid pace ever since the Industrial Revolution. The latter part of the 20th century, in particular, has seen a remarkable advance in computing technology. Two laws describing the advance of computing technology discussed below promise an unprecedented growth of AI.

10.5.1 Moore's Law

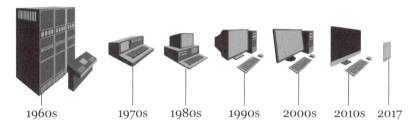

1960s   1970s   1980s   1990s   2000s   2010s   2017

Fig. 10.3 **Moore's Law**

Some of the experts in the semiconductor industry made a very important observation that there is a regular pattern in the increase of computing power with time. The result is summed up as Moore's law which states that the number of transistors on a silicon chip doubles every eighteen months (Fig. 10.3). The increase is not only in the number of transistors, but also in their operational speed. Effectively, the computing power of chips doubles every eighteen months. This computational doubling is observed in workstations, desktops, laptops, palmtops, tablets and smartphones.

10.5.2 Law of accelerating returns

According to the law of accelerating returns, technology on the whole develops not linearly, but exponentially. This law can be applied directly to the development of brain power as shown in Fig. 10.4. Current computing power is of the order of 1000 MIPS (million operations per second), while the human brain runs at several million MIPS. Simulating intelligent phenomena is therefore not yet possible with the low computing power of modern computers. The foreseen exponential growth in computing technology will revolutionize desktop computing power. Reliable projection for the development of an average human-like brain is around 2030, while that of the whole human race is around 2050.

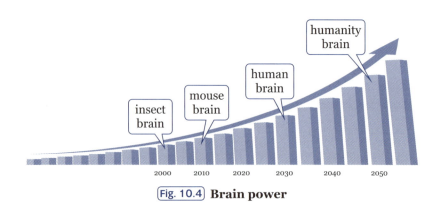

(Fig. 10.4) **Brain power**

10.5.3 AI growth stages

Although the limit to silicon-based intelligence is already in sight, Moore's law will continue metamorphosed empowered by nano-technology, DNA computing, Quantum computing and other genres of computing not yet on the horizon. These future technologies will continue advancing unbridled following the law of accelerating returns and become the launching pad for further development of AI. Experts predict the following significant stages of AI growth.

Artificial Narrow Intelligence (ANI)

The weak AI that we have seen in chapter 1 is also called "narrow," because it functions only in very narrow and limited domains. It cannot be extended to tasks for which it was not designed. *Deep Blue*, *Watson*, and *AlphaGo*, impressive as they are in outplaying human world champions, are outstanding examples of Artificial Narrow Intelligence.

Artificial General Intelligence (AGI)

The multi-faceted *strong* AI described in chapter 1 is also called general AI. Its capabilities will not be inferior to ours. It will be flexible enough to understand natural language, including puns and jokes, and interpret it in the appropriate context. The ability of the future Artificial General Intelligence to handle imprecise and ambiguous human language, coupled with common sense and wit, will make it indistinguishable from that of a human - taking it well beyond all imaginable versions of the Turing Test.

Artificial Superintelligence (ASI)

The extremely advanced form of future AI, called Artificial Superintelligence is predicted to become immeasurably more intelligent than the combined intelligence of the human species. Distributed on a network of cosmic proportions, and continuously self-developing and self-replicating, it will be ubiquitous and omniscient.

10.6 The approaching Singularity

In astrophysics, a *Singularity* refers to the center of a black hole where infinite mass is concentrated in a single point. Currently, no laws of physics can describe the dynamics of a *Singularity*. AI theorists suggest that AI with its exponential development is fast approaching a singularity-like phenomenon.

10.6.1 Singularity

Predicted by Einstein's General Theory of Relativity, a black hole is a region in space-time where the gravitational field is so strong that nothing, not even light photons, can escape from its grasp. The laws of physics can explain the dynamics of the back hole up to the event horizon (the boundary of no return). The center of the black hole, a point of zero volume and infinite mass, where laws of (classical) physics break down is called the *Singularity*.

Will the progress of AI eventually lead to an AI-Singularity? Fig. 10.5 shows the developmental stages of AI. ANI is expected to develop into AGI, which will be on a par with human intelligence. AGI

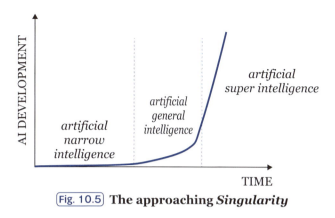

(Fig. 10.5) **The approaching *Singularity***

will have the power to improve itself over and above the human ini-
tiative. What follows next is anybody's guess. In the AI community
itself, expert opinion is divided. The conservative opinion is that
AGI is the goal of all AI projects. With AGI, we will have reached the
goal of "building machines that think and behave like human beings," the goal set by the AI founding fathers.

The radical opinion, on the other hand, basing itself on the law of
accelerating returns, states that AGI will form a self-improvement
feedback loop and continue self-development, with or without human intervention, until at some point before humans realize, it will
become full-blown Superintelligence. The point at which AGI gives
birth to ASI will be the *Singularity*.

But just when and how will the *Singularity* come about? The time-
line is still hotly debated in the AI community. The transition from
AGI to ASI, generally known as the takeoff, may be postulated as

any one of the scenarios described below.

Soft takeoff

The gradual development of AGI into ASI is called soft takeoff. The time-scale needed for the gradual transformation might be several decades or even centuries. For instance, some firms could create ASI in a highly controlled environment, experimenting with it, and tuning the initial parameters. The greatest concern in this area of research would be to lay the foundation for a sound ethical code in the "DNA" of software modules that will become the building blocks of ASI.

Hard takeoff

Another possibility is that AGI might suddenly reach a critical mass and get out of control, rapidly spreading over the internet like wildfire, making zillions of copies of itself and in no time develop into an omniscient and ubiquitous networked ASI. This scenario is called the hard takeoff. At present, this scenario appears to be hypothetical and more like the visions of sci-fi, but this future reality may be closer than we can imagine.

Aborted takeoff

In the quantum mechanical world, nature puts limits on the fundamental properties of matter and energy so much so that some bizarre quantum effects defy our classical (common sense) way of thinking. In the macroscopic world, too, there are limits. The speed of light is the universal limit beyond which no material body or information can travel. No amount of energy supplied to moving bodies and phenomena can accelerate them past the value of c (3×10^{8} m/sec). Moving bodies exhibit relativistic effects like *time dilation* and *length contraction* when approaching the speed of light.

Intelligence, too, might have a natural limit of which we are not yet aware.[2] AI will, no doubt, continue to grow for a long period of time; nevertheless, it could hit a barrier beyond which the law of accelerating returns might have to yield to the law of "diminishing returns." We cannot take it for granted that there will be no quantum mechanical and/or relativistic effects on information storage, retrieval, sharing, and processing when computational technology reaches a critical mass. These effects could put a hard limit on computationally produced intelligence, not allowing AGI to develop into ASI.

10.6.2 Humanity and Artificial Superintelligence

In the distant future, will humans and ASI co-operate to create a mutually self-fulfilling culture and society? What experts prophesy varies substantially in quantity and quality. The media portrays doomsday scenarios side by side with blissful paradise scenarios. The concern expressed by some of the most avant-garde thinkers of our times will provide food for thought to the serious AI researcher.

---

2   The author prefers to suggest the term "aborted takeoff" to refer to the limiting hypothesis of intelligence.

Unfriendly ASI

Hard takeoff may result in an uncontrolled and unfriendly ASI. Being orders of magnitude higher in intelligence than the most capable human minds, ASI will scoff at our human endeavors and as some AI critics observe rather sarcastically, "will keep us as pets." ASI may not have the "baser" human instincts built in, and thus would not need moral restraints. But it could be programmed with baser instincts by groups bent on achieving their own selfish goals and gaining an upper hand, for example, in the arms race.

The history of technology is replete with inventions that were developed primarily for military purposes. Computers and Internet which seemingly work for the service and welfare of humanity had their origins in military research. Many thinkers are concerned that ASI could be harmful to us, not because it is malicious or willfully against us, but because its goals may not correspond to the goals of human society. Human existence will be in danger if it tries to block ASI programmed to achieve its goals unthwarted.

Benevolent ASI

AI enthusiasts say that we need not worry because ASI will have no negative aspects in the post *Singularity* era. Their foresight rests on the basic assumption that an increase in intelligence is concomitant with an increase in wisdom. They say that the more we know, the wiser we become; and the wiser we are, the more ethically responsible we become. Crime and evil spring from our self-centeredness and pettiness which are rooted in biological and psychological needs. It is ignorance that blinds us to the universal good. Free from such needs, ASI will be benign and altruistic. It will judge its actions from the vantage point of universal good. AI enthusiasts are convinced that no problem will remain unsolvable for ASI. It will provide viable solutions to all sorts of problems known to humanity: poverty, hunger, misery and even death.

10.7 Exercises

1. Can you think of some more "objections" to machine intelligence?

2. How would you critique Asimov's laws of robotics?

3. Over-mechanization and over-robotization will rob us of our jobs and cause mass unemployment. What do you think?

4. Is passing the Turing Test a necessary and sufficient requirement for recognizing machine intelligence?

5. Is *Watson* capable of passing the Turing Test?

6. Give your reasons for or against the possibility of soft, hard and aborted take off.

7. Many experts warn that AI will soon get out of control and humanity will be in danger. What would you suggest to keep AI development in check?

8. Will the future AGI (if it actually materializes) be capable of explaining the mysteries of our lives and of the universe in which we live?

9. Describe the human-AGI interacting environment of the dis-

tant future.

10. Will AGI have consciousness and self-awareness like us?

References

Armstrong, S. (2014). *Smarter Than Us: The Rise of Machine Intelligence*. Machine Intelligence Research Institute.

Asimov, I. (1970). *The Rest of the Robots*. Panther.

Austin, J. L. (2016). *Emotional Intelligence Mastery: Why EQ can Often Matter More Than IQ*. CreateSpace Independent Publishing Platform.

Barrat, J. (2015). *Our Final Invention: Artificial Intelligence and the End of the Human Era*. St. Martin's Griffin.

Bostrom, N. (2016). *Superintelligence: Paths, Dangers, Strategies*. Oxford University Press.

Epstein. R., Roberts, G., & Beber, G. (Eds.). (2009). *Parsing the Turing Test: Philosophical and Methodological Issues in the Quest for the Thinking Computer*. Springer.

Kurzweil, R. (2000). *The Age of Spiritual Machines: When Computers Exceed Human Intelligence*. Penguin Books.

Kurzweil, R. (2006). *The Singularity Is Near: When Humans Transcend Biology*. Penguin Books.

Levesque, H. J. (2017). *Common Sense, the Turing Test, and the Quest for Real AI*. MIT Press.

Moravec, H. (2000). *Robot: Mere Machine to Transcendent Mind*. Oxford University Press.

AI GLOSSARY

agent: A software program capable of autonomous reasoning and action to achieve goals.

alpha-beta pruning: The threshold values in the Minimax algorithm that enable unpromising branches to be identified and eliminated from the search.

Ant Colony Optimization (ACO): A meta-heuristic optimization algorithm based on the foraging behavior of ants.

antecedent: The left-hand side of an *If-Then-Because* rule.

back propagation: In neural networks, the adjustment of the connection weights based on the training data with known output, thus training the network to eventually predict the output for unknown data.

backward chaining: A strategy of working backward to prove a goal.

blind search: A search conducted without any heuristic information.

bot: An AI agent working autonomously in the Web environment.

branching factor: The number of children of a given node in a search tree.

breadth first: A search strategy in which all the nodes at a given level are examined before proceeding to the next level.

Church-Turing Thesis: The proposition that computers can perform any specified symbolic process.

CLIPS: A forward-chaining Expert System shell using the C programming language.

combinatorial explosion: The exponential growth of the number of nodes to be explored in a search.

Computational Intelligence: Several nature-inspired computational methodologies for solving complex problems. It includes Fuzzy Logic, Neural Networks and Evolutionary Algorithms.

data mining: The process of extracting potentially useful information from large datasets.

deduction: The process of deriving new facts or theorems from known facts and axioms.

DENDRAL: An early Expert System that determined molecular structure of organic compounds given mass spectrometer data.

Emotional Intelligence (EI): The ability to identify, assess, and manage the emotions of oneself and those of others.

Expert System (ES): A software system capable of giving expert advice to end users in a given domain.

forward chaining: A data-driven strategy of working forward to the conclusion of a problem.

fuzzy set theory: A variant of set theory in which set membership is represented by the degree of membership ranging from 0 to 1.

game tree: A search tree for a game in which nodes represent the game states and arcs represent moves made by the players.

Genetic Algorithm (GA): An algorithm that computationally mimics the mechanisms and processes of biological evolution.

heuristics: The knowledge rules derived from the intuition and experience of domain experts. (It also refers to a *rough estimate*.)

hill climbing: A form of search in which the path of steepest ascent towards the goal is taken at each step.

hybrid algorithm: A combination of two or more algorithms designed to improve the performance of individual algorithms.

inference rule: A rule that combines facts to produce new facts or conclusions.

knowledge acquisition bottleneck: The fact that knowledge acquisition is the most difficult and lengthy step in developing an Expert System.

knowledge base: A collection of knowledge in an Expert System, including facts and rules.

knowledge engineer: A person who plays the key role in developing an Expert System. His job includes gathering knowledge from domain experts, organizing and codifying the knowledge.

knowledge engineering: The science and engineering of making Expert Systems.

Knowledge-base System (KBS): An alternative term for Expert System.

machine intelligence: An alternative term for Artificial Intelligence.

machine learning: An AI sub-discipline focused on having machines act without being programmed to do so. Machines learn the patterns in the data and adjust their behavior accordingly.

meta-heuristic algorithm: An optimization algorithm that can be applied to almost any type of optimization problem, since it does not rely on the heuristics of the problem domain.

minimax: An algorithm to perform search in a game tree to determine the best move.

Monte Carlo Tree Search (MCTS): A heuristic search algorithm used in searching for the winning moves in a game.

Natural Language Processing (NLP): The ability of computers to understand, or process natural human languages and derive meaning from them.

ontology: A set of objects and their relationships in a given domain. (In philosophy, ontology refers to the study of the existence of things.)

optimization: finding the (global) minima or maxima of functions, subject to a set of constraints.

Particle Swarm Optimization (PSO): A meta-heuristic optimization algorithm based on the flocking behavior of birds, bees, etc.

perceptron: An early artificial neuron.

planning: AI sub-discipline dealing with planned sequences or strategies to be performed by an autonomous agent.

PROLOG: Expert Systems Programming Language.

pruning: The use of a search algorithm to cut off undesirable solutions to a problem during the search.

Pseudocode: A program code written without using any particular programming language.

robot: A piece of hardware designed and programmed to carry on specific activities.

rule: The *If-Then-Because* format of representing knowledge in an Expert System.

search tree: A graph representing the moves and states of a game.

semantic network: The representation of meaning based on a network of nodes and labeled arcs. The nodes represent objects, while the arcs represent relations among the objects.

semantic web: The web formed by semantically structured information which is machine-readable.

semantics: The study of how language statements denote meanings.

shell: A software containing ready-made inference engine and user-interface of an Expert System.

Singularity: The explosion of Artificial Intelligence from *General* Intelligence into *Super* Intelligence.

strong AI: An area of AI development that seeks to make AI systems with capabilities very close to that of the human mind.

sub-goal: A goal that contributes to the accomplishment of a higher-level goal.

supervised learning: The training of a program to classify data by providing their correct classifications prior to learning.

Swarm Intelligence (SI): The collective behavior of decentralized, self-organized systems, natural or artificial.

synapse: The interface between the axon terminals of one neuron and the dendrites of another neuron.

Turing Test (TT): The test proposed by Allan Turing to test the intelligence of machines.

unsupervised learning: The training of a program to classify data without knowing the classes a priori. It is also called *clustering*.

weak AI: A computer system that operates within a predetermined range of skills, usually focused on a particular task in a domain. It is also known as *narrow* AI.

web content mining: The process of extracting useful information from the contents of Web documents.

web mining: The process of discovering previously unknown and potentially useful information from Web data.

web structure mining: The process of extracting useful structure information from the Web.

web usage mining: The process of extracting useful usage patterns from Web data.

weights: The connection strength between neurons in a neural network. These weights are adjusted in the process of learning.

wetware: The human brain, viewed as a kind of software.

INDEX

Artificial Intelligence:

A Non-Technical Introduction

2017年 3 月30日　第 1 版第 1 刷発行

著　者：ゴンサルベス　タッド
発行者：髙　祖　敏　明
発　行：Sophia University Press
　　　　上 智 大 学 出 版

　　　〒102-8554　東京都千代田区紀尾井町7-1
　　　URL：http://www.sophia.ac.jp/

　　　　　　制作・発売　㈱ぎょうせい
　　　〒136-8575　東京都江東区新木場1-18-11
　　　TEL 03-6892-6666　FAX 03-6892-6925
　　　フリーコール　0120-953-431
　　　〈検印省略〉　URL：https://gyosei.jp

印刷・製本　ぎょうせいデジタル㈱
ISBN978-4-324-10260-2
(5300264-00-000)
［略号：(上智)AI］
NDC分類007.13

Sophia University Press

　上智大学は、その基本理念の一つとして、
「本学は、その特色を活かして、キリスト教とその文化を
研究する機会を提供する。これと同時に、思想の多様性を
認め、各種の思想の学問的研究を奨励する」と謳っている。
　大学は、この学問的成果を学術書として発表する「独
自の場」を保有することが望まれる。どのような学問的
成果を世に発信しうるかは、その大学の学問的水準・評
価と深く関わりを持つ。
　上智大学は、(1) 高度な水準にある学術書、(2) キリス
ト教ヒューマニズムに関連する優れた作品、(3) 啓蒙的問
題提起の書、(4) 学問研究への導入となる特色ある教科書
等、個人の研究のみならず、共同の研究成果を刊行する
ことによって、文化の創造に寄与し、大学の発展とその
歴史に貢献する。